NATURA

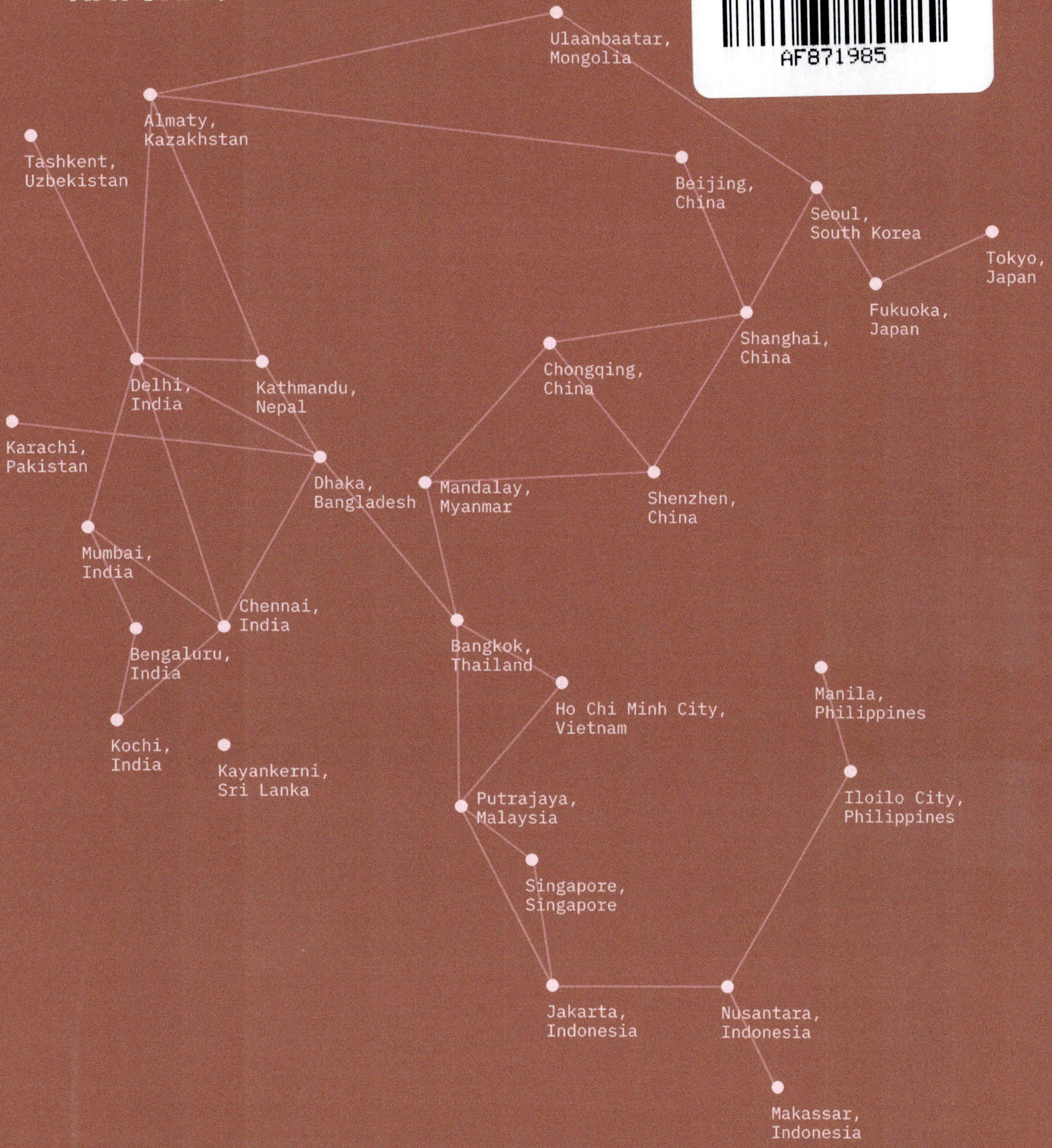

Asia Roadmap for Urban Nature-based Solutions

NYU | Global Access Publications

An imprint of
NEW YORK UNIVERSITY PRESS
New York
www.nyupress.org

Please contact the Library of Congress for Cataloging-in-Publication data.

ISBN: 9781479850402 (paperback)
ISBN: 9781479850426 (library ebook)
ISBN: 9781479850419 (consumer ebook)

This book is printed on acid-free paper, and its binding materials are chosen for strength and durability. We strive to use environmentally responsible suppliers and materials to the greatest extent possible in publishing our books.

The manufacturer's authorized representative in the EU for product safety is Mare Nostrum Group B.V., Doelen 72, 4831 GR Breda, The Netherlands.
Email: gpsr@mare-nostrum.co.uk.

Manufactured in the United States of America
10 9 8 7 6 5 4 3 2 1
Also available as an ebook

Design & Layout by Phuong Anh Bui

Lead Authors

Maitreyi Koduganti
Indian Institute for Human Settlements
University of Illinois - Urbana Champaign

Erich Wolff
Nanyang Technological University
Utrecht University

Chandni Singh
Indian Institute for Human Settlements

Perrine Hamel
Nanyang Technological University

Contributing Authors

Yeowon Kim
Korea University

Sheetal Patil
Indian Institute for Human Settlements

Alex M. Lechner
Northeastern University China

Pamela Cajilig
University of the Philippines

Loan Diep
New York University
The New School

Timon McPhearson
New York University
The New School

This report should be cited as:

Koduganti, M., Wolff, E., Singh, C., Hamel, P., Kim, Y., Patil, S., Lechner, A.M., Cajilig, P., Diep, L., & McPhearson, T. (2026) *Asia Roadmap for Urban Nature-Based Solutions*. In: The Global Roadmap for Urban Nature-Based Solutions. New York City, NY: NATURA. https://doi.org/10.18574/nyu/9781479850426.0010001.

Contributions and Acknowledgements

Maitreyi Koduganti, Erich Wolff, Chandni Singh and Perrine Hamel led data synthesis and writing of the Regional Report, with contributions from Yeowon Kim, Sheetal Patil, Alex M Lechner, Pamela Cajilig, Loan Diep, and Timon McPhearson. All authors contributed to the final recommendations and reviewed the report for accuracy and clarity. The authors thank Sheetal Patil, Yeowon Kim, Alex M. Lechner, Weiqi Zhou and Meherwan Patel for reviewing the draft and providing critical inputs. The authors thank Ho Huu Loc, Baolong Han, Pamela Cajilig, Alex M. Lechner, Ravi Jambhekar, Jagdish Krishnaswamy, Divyanshi Vyas, Hammad Gilani, Jenia Mukherjee, Mihir R. Kulkarni, Madhuri Ramesh, Jahin Sakkhar, and Yeowon Kim for their valuable contribution in developing and writing the case studies that informed and supported the findings of this report.

The study was supported by Natalie Pierson (research and coordination), Rebecca Tiernan, Vanya Bisht, and Dave Kendal (systematic mapping), Sofya Krasnaya and Raffaello Khan Agustin (review of reviews), Carole Alazki (case study data analysis), Veronica Olivotto (case study review and copy editing), Phuong Anh Bui and Valérie Lechêne (design). Timon McPhearson and Loan Diep conceptualized and led the overall Global Roadmap initiative, including regional report structure, design and final editorial review. The NATURA 'Global Roadmap for Urban Nature-based Solutions' was funded by the National Science Foundation.

Co-Leads

Timon McPhearson
New York University
The New School

Loan Diep
New York University
The New School

Advisors

Niki Frantzeskaki
University of Utrecht

Erik Andersson
University of Helsinki

Rob McDonald
The Nature Conservancy

Pippin Anderson
University of Cape Town

Harriet Bulkeley
Durham University

Yeowon Kim
Korea University

Weiqi Zhou
Chinese Academy of Sciences

Literature Review

Loan Diep
New York University
The New School

Rebecca Tiernan
Arizona State University

Natalie Pierson
New York University
The New School

Vanya Bisht
University of Waterloo

Dave Kendal
Future in Nature

Case Study Analysis

Carole Alazki
New York University
The New School

Natalie Pierson
New York University
The New School

Veronica Olivotto
New York University
The New School

Copy Editing

Veronica Olivotto
New York University
The New School

Design

Phuong Anh Bui
New York University
The New School

Funding

This report has been primarily funded by the US National Science Foundation Accel-Net program to the NATURA Project (Grants #1927468 and #1927167). This project has received additional support from the National Research Foundation Singapore, Prime Minister's Office, Singapore (award NRF-NRFF12-2020-0009), the CO-CARBON project, funded by the Strategic Research Council (SRC) within the Research Council of Finland (grant nos. 358254 and 358256), the SMARTer Greener Cities project through the Nordforsk Sustainable Urban Development and Smart Cities program (Project no. 95377), and the PLANNING with YOUTH project (grant no. 2019-01887) funded by FORMAS - the Swedish Research Council for Sustainable Development under the targeted call National Research Program for Sustainable Spatial Planning. The authors also acknowledge support from Agence Française de Développement (AFD) through the "Greening Urban Food Systems: Building Sustainable urban agriculture practices in Bengaluru through nature-based solutions' project (Grant No. AFD/ISR/ECO/771-2022).

Table of Contents

Executive Summary

Asia is home to over 2.3 billion urban residents and is expected to grow another 50% by 2050. The way nature is included in urban growth will play a decisive role in shaping urban futures around the world.

Singapore's Bishan-Ang Mo Kio Park. Erich Wolff, 2025.

Asia is urbanizing at an unprecedented pace and is home to over 54 percent of the world's urban population. Shaped by highly diverse socio-demographic, political, and ecological and climatic contexts, this rapid expansion will continue in the coming decades, presenting significant challenges in the face of climate change and mounting social and political pressures. Megacities such as Tokyo (37.8M), Jakarta (35.4M), and New Delhi (34.7M) are at the forefront of global urbanization, and by 2050, Asia's population is projected to see a 50 percent rise in its urban population (UN ESCAP, 2025). At the same time, many secondary and smaller cities—often at the urban–peri-urban–rural interface and with large informal settlements—are expanding rapidly with limited planning capacity.

Most urban areas in Asia are shaped by monsoon, deltaic, or coastal conditions, making water management and heat mitigation central urban challenges. Around 411 million people across 330 Asian cities (each with more than 300,000 residents), including megacities like Shanghai, Karachi, and Kolkata, are currently exposed to drought risks. Further, many Asian cities are also grappling with increased flooding and intense heatwaves—climate change impacts that disproportionately affect vulnerable populations. These pressures intersect with entrenched inequalities, insecure land tenure, and uneven access to basic services, particularly in informal settlements.

In response to these escalating threats, Nature-based Solutions (NbS) have emerged as a vital strategy in urban and urbanizing spaces to address climate impacts, as well as environmental degradation, water scarcity, and public health risks. Across the region, NbS principles are adopted under diverse and overlapping concepts, such as sponge city, blue–green infrastructure, water-sensitive urban design, *satoyama* landscapes, and other culturally rooted practices, showing how NbS exist within distinct socio-ecological and cultural contexts, even as varied terminology can complicate comparison and coordination. NbS hold particular promise for enhancing resilience, sustainability, and equity, but their benefits and burdens are not yet evenly distributed.

Drawing on a systematic review of academic literature, grey literature and 27 case studies from 26 cities of 16 countries across the region, the *Asia Roadmap for Urban Nature-based Solutions* synthesizes current trends of NbS practices and policies, governance models, financial mechanisms, challenges and opportunities for advancing NbS across urban centers in South, Southeast, Central, and East Asia. It highlights how global agendas and international funders interact with local priorities, and how evidence gaps in monitoring and evaluation constrain learning and scaling.

KEY MESSAGE 1:

Uptake of urban NbS in Asia is blooming, but adoption is uneven across subregions. Countries including China, Singapore, Japan, and South Korea are leading in large-scale implementation, yet diverse NbS projects continue to emerge across South, East and Southeast Asian cities.

Key findings: Urban NbS in Asia range from national programs in China and Singapore, flagship urban renewals like Seoul's Cheonggyecheon Park, to community-led initiatives in informal settlements and peri-urban areas across.

India and Indonesia. While East, South, and parts of Southeast Asia exhibit strong experimentation, Central Asia shows limited uptake and sparse documentation. NbS extend across multiple ecosystems—from urban forests and parks to mangroves, wetlands, and river corridors—in climates spanning tropical deltas to arid inland cities. However, uptake remains limited in regions such as Central Asia, where urbanization rates are lower and documentation is sparse. Many locally rooted practices remain under-recognized because they are not labeled as "NbS," are described only in local languages, or lack systematic monitoring and reporting

Key recommendations: Establish regional databases and cross-learning platforms to capture locally grounded NbS—including those not explicitly named as such —and share experiential knowledge on implementation and governance. These efforts can support monitoring, scaling, and mainstreaming NbS across cities and regions, especially those underrepresented. Improved knowledge exchange is critical to demonstrate workable solutions, inspire new initiatives, and support city-to-city learning.

KEY MESSAGE 2:

Urban NbS are proving to be effective in addressing climate and environmental risks in Asian cities, yet the full range of benefits remains unequally distributed. Most projects prioritize reducing risks from flood and extreme heat; and improving water security, air quality, and biodiversity, but long-term social benefits receive relatively lower attention.

Key findings: Urban NbS are predominantly used to address specific climate adaptation challenges, especially urban flooding and extreme heat, which are major risks in countries such as Thailand, the Philippines, South Korea, and India. This focus has driven strong investment in blue–green infrastructure, including several flagship projects of international visibility. However, many important co-benefits—such as livelihood generation, mental health improvements, food security, cultural values, and soil quality enhancement—remain under-evaluated or inconsistently measured across cities. As a result, the current evidence base provides only a partial picture of how benefits are distributed, who gains or is excluded, and where trade-offs or disservices may occur.

Key recommendations: Develop systematic, context-sensitive monitoring, evaluation, and learning frameworks that track both biophysical performance and social outcomes, including distributional equity and potential disservices. Metrics should aim to capture the full range of NbS benefits—from biodiversity gains and hazard reduction to health, livelihoods, and cultural values—and explicitly assess risks of displacement or exclusion.

KEY MESSAGE 3:

Urban NbS in Asia are largely driven by national or municipal governments or led by communities. While multilaterals, NGOs, and private actors play important roles in NbS implementation, they are proportionally less active. Stronger partnerships are needed to bridge top-down and bottom-up approaches, enhance durability, and enable long-term scaling.

Key findings: Urban NbS in Asia are led by a wide range of actors, however, they translate primarily into practice through two main models: government and municipality-led programs—such as China's Sponge City initiative and Singapore's river restoration efforts; and community-driven initiatives, including urban agriculture and neighborhood greening in India and Indonesia. While these models demonstrate strong leadership, effective long-term implementation requires deeper collaboration across governments, civil society, academia, and the private sector. Promising examples, such as the multi-actor governance of the East Kolkata Wetlands, show how coordinated partnerships can enhance ecological restoration and community benefits. However, such integrated arrangements remain the exception rather than the norm.

Key recommendations: Foster multi-stakeholder coalitions, including carefully designed public-private-people partnerships, to catalyze context-specific innovation. Academic and research institutions should play a key role in evidence generation, co-design and the development of evaluation frameworks. Long-term collaboration mechanisms are needed to align roles, responsibilities, and resources across actors and scales.

KEY MESSAGE 4:

Equitable urban NbS in Asia depend on participatory and locally led approaches, especially in informal settlements. Asia is home to 60% of the world's informal settlements population, and NbS in these areas require place-based and tailored support.

Key findings: Examples such as the RISE Program in Indonesia and small-city NbS initiatives exemplify how co-designed approaches can improve water, health, and resilience outcomes, especially within informal and rapidly urbanizing areas. However, many NbS initiatives continue to rely on top-down approaches, state-led models that often overlook local voices and priorities. Insecure land tenure, limited space, and weaker participation frameworks often hinder adoption within informal settlements, despite these areas standing to gain significantly.

Key recommendations: Institutionalize participatory urban planning, gender-sensitive design, and decentralized governance models early-on within project cycles. Provide resources for long-term engagement, co-design, and community stewardship, rather than short, project-bound consultation. Support locally led experiments and recognize community-driven NbS (e.g., urban farming, household-level water harvesting) as part of official adaptation strategies, not as peripheral practices.

KEY MESSAGE 5:

Finance for urban NbS in Asia is expanding, yet investments remain concentrated in large projects, leaving community-based NbS underfunded. Diversifying finance through blended models—combining public budgets, concessional loans, green-blue bonds, and results-based or pooled funds—can help mainstream urban NbS more effectively.

Key findings: Urban NbS in Asia rely on a mix of financing instruments—including public budgets, short-term grants, concessional loans, public–private partnerships, and emerging tools such as green and blue bonds or ecosystem credit schemes—but access to these resources varies sharply across city contexts and project scales. Large, state-led NbS projects are more likely to secure national or multilateral funding, while smaller municipal and community initiatives depend heavily on project-bound grants that undermine continuity and long-term maintenance. Although the region has a diverse funding landscape, the difficulty of monetizing adaptation and other public-good benefits limits private-sector engagement, reinforcing reliance on a narrow group of public and development funders and contributing to persistent financing gaps.

Key recommendations: Develop blended finance models that combine public funds, concessional loans, and private capital, including municipal green bonds, micro-grants, and community co-financing where appropriate. Create a menu of financeable NbS options tailored by city typology (e.g., climate zone, population size, and urban density) to guide funders and governments in making scalable investments. Strengthen cost–benefit frameworks and evidence on long-term co-benefits to de-risk NbS for investors and public agencies.

KEY MESSAGE 6:

NbS in Asian cities are not yet systematically integrated into urban planning and climate action agendas, with fragmented governance limiting mainstreaming efforts. While national and city-level frameworks such as China's Sponge City policy, Singapore's City in Nature, and India's Smart Cities Mission illustrate forms of NbS mainstreaming, weak legal instruments, siloed institutions, and limited monitoring, evaluation, and learning systems constrain consistent implementation.

Key findings: Governance of urban NbS in Asian cities remains fragmented, with most initiatives led by individual government bodies and limited mechanisms for cross-sectoral coordination. In countries like Thailand, Indonesia, and India, greening strategies are often not systematically embedded across sectors such as housing, transport, public health, cultural heritage, or tourism. Fragmented responsibilities, limited technical capacity, and siloed sectoral approaches often weaken coordinated implementation. While centralized, state-led models like China's Ecological Civilisation model can accelerate progress, they risk overlooking local priorities—especially when cross-sectoral collaboration is weak and governance across national, regional, and local levels is poorly aligned.

Key recommendations: Strengthen institutional coherence by embedding NbS into core urban planning, land-use, infrastructure and other development policies. Align NbS with national and local governance strategies, enabling their integration into broader climate adaptation and sustainable development agendas as well as creating beyond-government collaborations. Collaboration across governance levels is crucial to align.

KEY MESSAGE 7:

Embedding Indigenous and local knowledge into urban planning can strengthen the cultural relevance, inclusivity, and long-term success of urban NbS. Asia is home to nearly 70% of the world's Indigenous Peoples, making Indigenous and locally rooted knowledge systems relevant to urban NbS design and governance.

Key findings: Practices such as Japan's *satoyama* landscapes, Myanmar's floating gardens, and Indonesia's Subak systems illustrate how context-sensitive traditional ecological knowledge enhances resilience and biocultural diversity.

Many urban and peri-urban NbS in Asia may draw on these knowledge systems in practice, however, such knowledge is often excluded or not recognized within formal planning. This disconnect limits the potential for NbS to support self-determination pathways, and locally meaningful solutions.

Key recommendations: Promote cross-sectoral collaboration that aligns Indigenous and local knowledge with urban planning processes. Build context sensitive and inclusive NbS frameworks that reflect diverse voices and worldviews, thus, ensuring cultural, ecological, and social relevance. Invest in capacity-building and long-term partnerships that move beyond symbolic consultation toward shared governance and co-management of urban nature.

List of Figures

List of Tables

Acronyms and Abbreviations

Acronyms and Abbreviations	Full Term/Description
Abbreviation	Full Name
ABC	Active, Beautiful, and Clean
ACCCRN	Asian Cities Climate Change Resilience Network
ACROS	Asian Cross Roads Over the Sea
ADB	Asian Development Bank
AFD	Agence Française de Développement (French Development Agency)
ASEAN	Association of Southeast Asian Nations
BDP	Bangladesh Delta Plan
BDT	Bangladeshi Taka
BMA	Bangkok Metropolitan Administration
BMU	German Federal Ministry for the Environment, Nature Conservation, and Nuclear Safety
CBO	Community-based Organizations
CDKN	Climate & Development Knowledge Network

Acronyms and Abbreviations	Full Term/Description
CIAL	Cochin International Airport Limited
CIA	Cochin International Airport
CMEMP	Coastal and Marine Ecosystems Management Program
CNRS	Centre for Natural Resource Studies
CNY	Chinese Yuan Renminbi
DBC	Delta Blue Carbon
Eco-DRR	Ecosystem-based Disaster Risk Reduction
EKW	East Kolkata Wetlands
EKWMA	East Kolkata Wetlands Management Authority
EU	European Union
EUR	Euro
GBF	Kunming-Montreal Global Biodiversity Framework
GCF	Green Climate Fund
GDP	Gross Domestic Product
GI	Green Infrastructure
GIZ	Deutsche Gesellschaft für Internationale Zusammenarbeit (German Agency for International Cooperation)
GREEN	Growth, Resilience, Energy Efficiency, and Nature-Based Solutions
IUCN	International Union for Conservation of Nature
INGO	International Non-Governmental Organization
INR	Indian Rupee
JBE	Japan Blue Economy Association
KFS	Korean Forest Service
KMC	Kolkata Municipal Corporation
LAC	Latin America and the Caribbean
LGU	Local Government Unit
LID	Low Impact Development
MEL	Monitoring, Evaluation and Learning
MENA	Middle East and North Africa
MET	Mongolian Ministry of Environment and Tourism
MLIT	Ministry of Land, Infrastructure, Transport and Tourism
MOCC	Ministry of Climate Change
MOE	Ministry of Environment

Acronyms and Abbreviations	Full Term/Description
MoEFCC	Ministry of Environment, Forests and Climate Change
MOF	Ministry of Finance
MOHURD	Ministry of Housing and Urban-Rural Development
MOC	Ministry of Construction
MONRE	Ministry of Natural Resources & Environment
MPA	Marine Protected Area
MWR	Ministry of Water Resources
NABARD	National Bank for Agriculture and Rural Development
NACFF	National Adaptation Fund for Climate Change
NATURA	Nature-based Solutions for Urban Resilience in the Anthropocene
NAP	National Adaptation Plan
NbS	Nature-based Solution
NBSAP	National Biodiversity Strategies and Action Plans
NDC	Nationally Determined Contribution
NGO	Non-Governmental Organization
NIUA	National Institute for Urban Affairs
NParks	National Parks Board
OECM	Other Effective Area-Based Conservation Measures
OECD	Organisation for Economic Co-operation and Development
PFI	Private Finance Initiative
PPP	Public-Private Partnership
PPJ	Perbadanan Putrajaya
PUPR	National Ministry of Public Works and Housing
PUB	Singapore’s National Water Agency
RISE	Revitalising Informal Settlements and their Environments
SFD	Sindh Forest Department
SDG	Sustainable Development Goal
SPV	Special Purpose Vehicles
SRC	Strategic Research Council
THB	Thai Baht
TNC	The Nature Conservancy
UNA	Urban Nature Atlas
UN	United Nations

Acronyms and Abbreviations	Full Term/Description
UNDP	United Nations Development Programme
UNECE	United Nations Economic Commission for Europe
UNESCO	The United Nations Educational, Scientific and Cultural Organization
UNHCR	United Nations Refugee Centre
UPA	Urban and Peri-Urban Agriculture
UPAGrI	Urban and Peri-Urban Agriculture as Green Infrastructure
USD	US Dollar
WISA	Wetlands International South Asia
WSUD	Water Sensitive Urban Design
WWF	World Wide Fund

1 | Introduction

Climate change is increasingly threatening livelihoods in Asia with significant risks from extreme weather events compounded by rapid land-use change and unplanned urbanization.

Himalayan cities such as Thimpu, Bhutan are seeing increasing built up area, with repercussions for green cover and exposure to hazards. Chandni Singh, 2018.

Asia covers one-third of the Earth's surface and almost every known climate occurs on this continent. For example, the annual mean surface air temperatures in the sub-region exhibit considerable variation, ranging from −3 °C in the coldest parts of North Asia to 25 °C in the warmest areas of Southeast Asia (Kobayashi et al., 2015). In North Asia, polar, subarctic, and temperate climates dominate, while Central Asia is characterized by arid and highland climates. South and Southeast Asia features a mix of grassland and tropical rainforest climates, whereas East Asia experiences a temperate climate that includes humid subtropical, humid continental, semi-arid, arid, and highland conditions. In Central Asia, semi-arid and desert climates prevail, reflecting the continent's vast geographical and climatic diversity. Asia's urban landscape is remarkably diverse, reflecting the continent's varied economic, cultural, and political contexts. Figure 1 presents a map of Asia and its countries for contextual reference.

Globally, climate change is a growing challenge and the stakes are particularly high for Asia. Asian urban areas face significant risks from projected climate change impacts, extreme weather events, unplanned urbanization, and rapid land-use change (Shaw et al., 2022). Average temperatures across the continent are rising at twice the global rate (Dabla-Norris, Nozaki and Daniel, 2021), accompanied by an increased frequency and intensity of extreme weather events such as flooding, heatwaves, and typhoons. Megacities like Dhaka and Bangkok are highly vulnerable to flooding driven by heavy rainfall and rising sea levels, posing severe threats to infrastructure and livelihoods (Dabla-Norris, Nozaki and Daniel, 2021; Cao et al., 2021). Heatwaves also pose a significant public health risk, particularly in densely populated urban areas such as Ho Chi Minh city, where the compounding effect of heat islands increases temperatures (Huang et al., 2020; Ramsay et al., 2021). Coastal cities like Jakarta, Singapore and Manila are

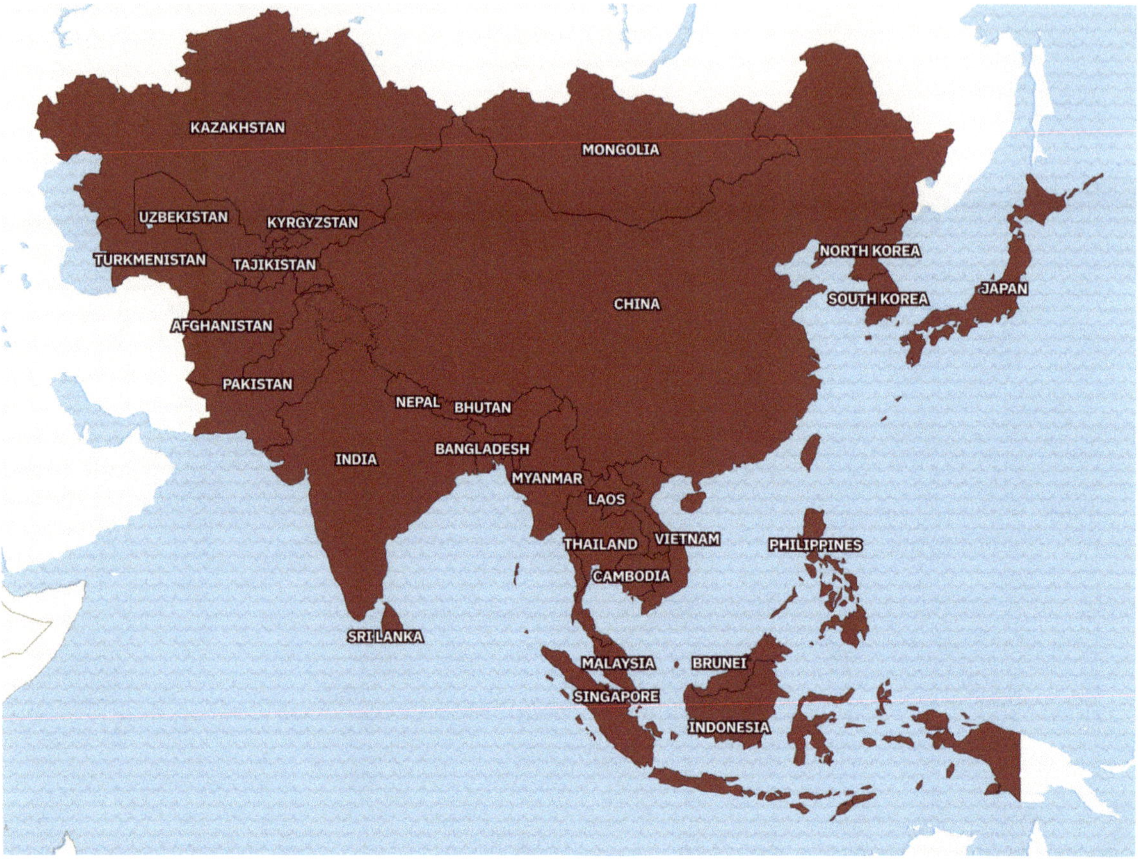

Figure 1: Asia's Regional Delineation Used in this Report. The boundaries and names shown, and the designations used on the map do not imply official endorsement or acceptance by NATURA

exposed to sea-level rise, threatening to displace millions of residents and disrupt local economies (Dodman et al., 2023). Central Asian cities are increasingly vulnerable to shifting precipitation patterns, which impact agricultural productivity and water availability. Yet, many questions remain unanswered regarding how population growth and urbanization affect such risks.

Asia is an incredibly populous continent, home to over half of the world's urban population, a share that is projected to rise to nearly two-thirds by 2050 (Shaw et al., 2022). Further, megacities around the world are projected to increase from 44 today to 67 in 2050 (Ghazi, 2024), with the largest increase being in Asia (Figure 2). Highly urbanized countries like Japan, Singapore and South Korea, have over 90 percent of their populations living in cities and are characterized by advanced infrastructure and aging populations (World Bank, 2023b). In contrast, lower middle-income countries like Cambodia and Nepal remain predominantly rural, with urbanization levels below 25 percent (United Nations, 2017). The rapid urbanization in China has seen over 60 percent of its population now residing in cities, a dramatic shift from 18 percent in 1978, driven by economic reforms and government policies (Hong et al., 2021). Meanwhile, India's urbanization, currently at 34 percent, is marked by significant regional disparities, with megacities like Mumbai and Delhi juxtaposed against slower-growing rural regions (Office of the Registrar General & Census Commissioner, India, 2011). Compared to cities in East and South Asia, Central Asian cities generally exhibit lower population densities, which highlights the region's diverse socio-economic conditions (World Bank, 2023a).

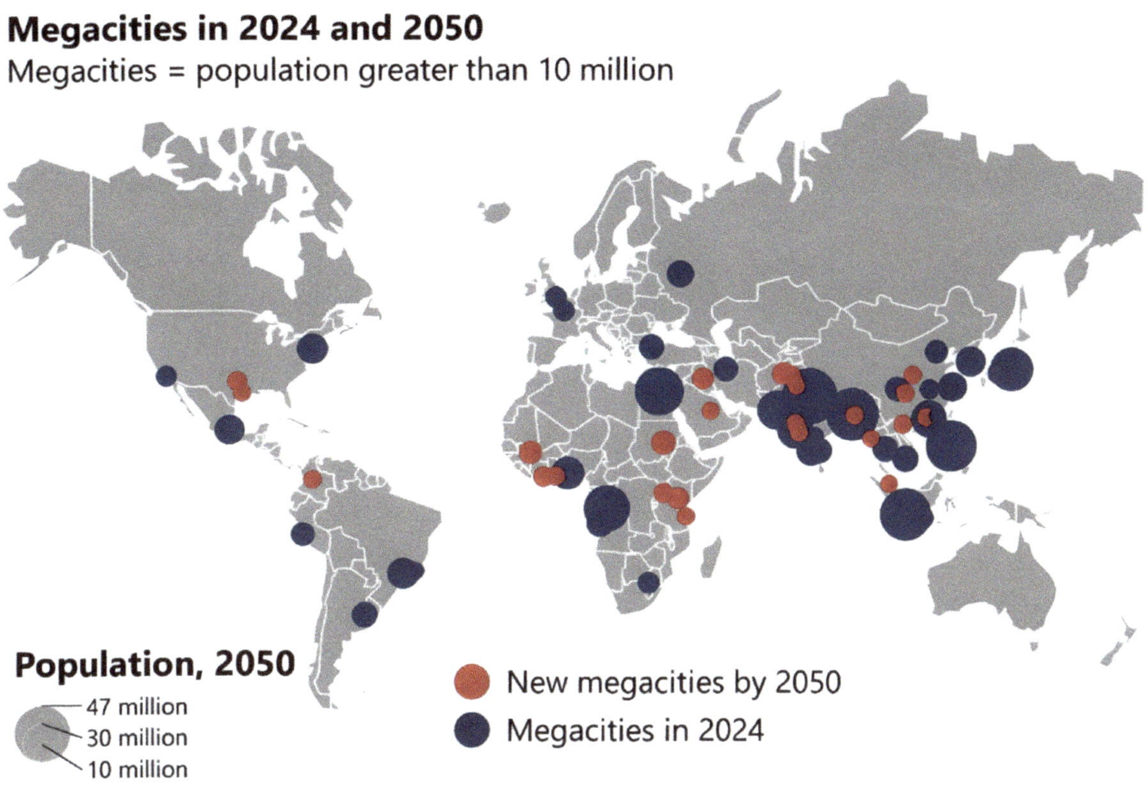

Figure 2: Rise of New Megacities, Globally. Oxford Economics Global Cities, 2023. https://www.oxfordeconomics.com/resource/rise-of-new-megacities-will-drive-global-urban-growth/

1.1 Urban Nature-based Solutions in Asia

Asian cities are increasingly adopting NbS to respond to the pressures of rapid urbanization while improving urban resilience to climate change.

In addition to climate change, key environmental challenges in the region include air and water pollution, waste management issues, and loss of biodiversity. For example, Delhi and Jakarta face severe air quality problems driven by vehicular emissions, industrial discharge, and construction dust (Firdaus, Elliott and Ibanez, 2023; Mollan, 2024). Water scarcity is another critical issue, particularly in Almaty, Tashkent, Chennai, and Jakarta, where over-extraction of groundwater and pollution of surface water further strain resources (Asian Development Bank, 2020; Singh, Jain, et al., 2021). Additionally, urban waste generation has surged, leading to significant pollution management challenges as many cities lack adequate infrastructure to manage increasing volumes effectively (Zhang et al., 2024).

Waste disposal in a Kampung (urban village) near Jakarta (Indonesia). Erich Wolff, 2019.

Waste disposal and accumulation along the Mutha riverbank in Pune (India). Prathijna Poonacha.

The pace of urbanization across Asia has often exceeded cities' capacity to provide adequate housing, services, decent employment, and infrastructure. This has led to a rise in urban poverty and widening social inequalities (Gadgil and Baker, 2017). Informal settlements have proliferated in cities like Manila and Mumbai, where up to half of the population have inadequate access to basic services such as clean water, sanitation, and healthcare. Overall, Asia is home to the largest population living in informal settlements globally—332 million people in Eastern and Southeast Asia and 197 million in Central and Southern Asia (Shaw et al., 2022). Although the proportion of urban residents living in informal settlements is gradually declining, the absolute number continues to rise. Across South Asia, Southeast Asia, and Central Asia, around 30 percent of the urban population resides in informal neighborhoods (United Nations Human Settlements Programme, 2020).

Meanwhile, many cities in East Asia—including Seoul, Tokyo, and Shanghai—are grappling with challenges related to rapid population densification. The combination of environmental pressures, social inequality, and infrastructure deficits underscore the need for integrated solutions. NbS offers the potential to address these challenges holistically, enhancing urban resilience while delivering social, economic, and ecological co-benefits. Notable examples of NbS policies include the 13th Five-Year Plan for China, which focuses on Sponge Cities (KPMG China, 2016); the strong promotion of water sensitive urban design in Vietnam (Asian Development Bank, 2019); and the East Kolkata Wetlands program in India that enables wastewater treatment and flood control.

Informal settlements in Bangkok (Thailand) are prone to flooding. Erich Wolff, 2022.

Informal settlement area in Bhopal (India) face heightened heat risk. Chandni Singh, 2025.

1.2 The State of Knowledge on Nature-based Solutions in Asia

NbS in Asian cities are increasingly recognized as key strategies for addressing interconnected urban environmental and societal challenges. After Europe, Asia accounts for the second highest volume of empirical research on urban NbS globally. Countries producing a large amount of research on environmental sciences and urban planning, such as China, Singapore and South Korea, are playing a key role in the development of technology and policy for NbS in Asia. Additionally, practices from large countries in East and South Asia have shaped how urban NbS are understood and applied across the region and have contributed to building a robust knowledge base for NbS in Asia (Luo et al., 2023).

China and India have led important developments in conventional and emerging forms of NbS in urban settings over the last decade. In China, cities such as Beijing and Shanghai are often highlighted as key sites for advancing knowledge on sponge cities – a concept that gained currency as governments and private developers rolled out green-blue-grey infrastructure management strategies to promote urban resilience over recent decades (Yin et al., 2022). In India, several government and civil society-led projects are building upon existing knowledge on ecosystem-based approaches and applying them in urban contexts to support the creation of diverse green infrastructure projects and broader urban transformation missions (Abbu et al., 2015). Additionally, significant investments in new technologies and approaches for climate adaptation, pollution remediation and resilience in these countries has supported a booming research landscape for NbS in Asia.

Other countries in East Asia including Japan and South Korea illustrate how NbS research and practice varies significantly based on the country and local culture. Japan's use of NbS in peri-urban landscapes is derived from techniques used in the *satoyama* system, a local reference for sustainable management of rural areas. Current projects advancing NbS in South Korea build upon an established knowledge base on urban river restoration practices prioritizing the connection of communities to revitalized waterways. In contrast, urban development in parts of Central Asia continue to prioritize conventional grey infrastructure developments over ecosystem-based alternatives (World Bank, 2023b). Slower development of NbS knowledge in these countries can also be attributed to the need to address immediate urbanization challenges over long-term sustainability goals.

Across the region, emerging research is showing that NbS can drive progress in the goals set by agreements such as the Kunming-Montreal Global Biodiversity Framework (GBF), the Paris Agreement and the Sustainable Development Goals (SDGs). The growing database of examples in the region indicates that NbS can be used to tackle the interconnected challenges of improving living conditions in urban areas, supporting climate adaptation and mitigation, and halting biodiversity loss in Asia (Hamel and Tan, 2022). Within debates about the National Biodiversity Strategies and Action Plans (NBSAP), China has been advancing the development of knowledge and frameworks on how to facilitate the reporting of progress in the goals of the GBF, notably through its state-led Ecological Civilization agenda (Zhu et al., 2024). In addition to international agreements, NbS knowledge production in Asia is also shaped by the financing landscape, in particular for low- and middle-income countries which often receive funding from multilateral banks.

These examples reflect how NbS knowledge in Asia draws on different traditions and precedents, responding to local social, ecological, geographic, economic, and political conditions. However, differences in languages and regional context contribute to a fragmented knowledge base on NbS. With few exceptions of initiatives to share knowledge among members of the Association of Southeast Asian Nations (ASEAN), the expansion of knowledge on NbS in the region is still partially constrained by national boundaries, being led by external partners, such as the European Union (Goethals, 2025). Despite these barriers, knowledge bases on NbS in Asia are being developed driven by regional and global ambitions to tackle environmental degradation and climate change impacts, which often disproportionately affect the urban poor, deepen inequalities, and limit access to essential resources (Padawangi, 2012; Satterthwaite et al., 2018; Dangol and Carrasco, 2019).

Constructed wetlands are used in many cities in Southeast Asia, including Putrajaya (Malaysia) and Singapore.
Erich Wolff, 2025.

1.3 Contributions of This Report

The *Asia Roadmap for Urban Nature-based Solutions* is one of seven regional assessment reports in the NATURA Global Roadmap for Urban Nature-based Solutions. It documents the practice of NbS in Asian cities and related concepts, examining the scale of interventions, funding mechanisms, key actors, and the challenges involved. This analysis lays the groundwork for further mainstreaming NbS in the region by providing a systematic overview of how Asian cities are adopting and adapting these approaches.

The report synthesizes academic and gray literature alongside case studies to illustrate how Asian cities are interpreting and implementing urban NbS. Examples include stream restorations, green corridors, urban parks, and various educational gardening programs that respond to local needs and policy frameworks targeting pollution reduction, climate change adaptation, and improved public health. The assessment also reviews established programs, such as China's large-scale sponge city investments, while also drawing attention to emerging trends in rapidly urbanizing contexts like Singapore and Bangkok, as well as NbS developments in more sparsely populated areas of Central Asia.

In addition, the report analyzes governance strategies and landmark projects spearheaded by different countries, showing how NbS have become a promising approach to address challenges such as informal urbanization and accelerated encroachment over peri-urban ecosystems (Mukhopadhyay and Chowdhury, 2021; Andrews et al., 2023). By compiling these diverse experiences, this report contributes to the global assessment by offering region-specific insights into the opportunities and barriers for mainstreaming NbS across different socio-political and ecological contexts.

2 Scope and Process

Learning spaces at the Hive, campus of the Nanyang Technological University (Singapore). Erich Wolff, 2024.

2.1 Overall Approach

This report is part of the NATURA Global Roadmap for Urban Nature-based Solutions—a coordinated initiative to synthesize the state of knowledge and practice across seven world regions (US-Canada, Latin America & the Caribbean, Europe, the Middle East & North Africa, Africa, Asia, Oceania). The methodology follows a framework used across the global and other regional assessments, enabling both cross-regional comparison and region-specific analysis. Within this shared framework, the Asia report offers a wide perspective on NbS in Asian cities. It highlights regional case studies to develop detailed profiles of NbS practices within policy, planning, and implementation frameworks. The approach encompasses three main steps: (1) exploring patterns and gaps in NbS research across the region; (2) highlighting case studies that reflect the practical applications of NbS in urban settings; and (3) synthesizing knowledge from this dual perspective, identify key knowledge gaps and highlight opportunities for scaling and adapting NbS to different Asian city contexts. This process was carried out through an iterative, collaborative and consensus-oriented co-production approach, drawing on regional expertise and global coordination to ensure both comparability and local relevance (Figure 3).

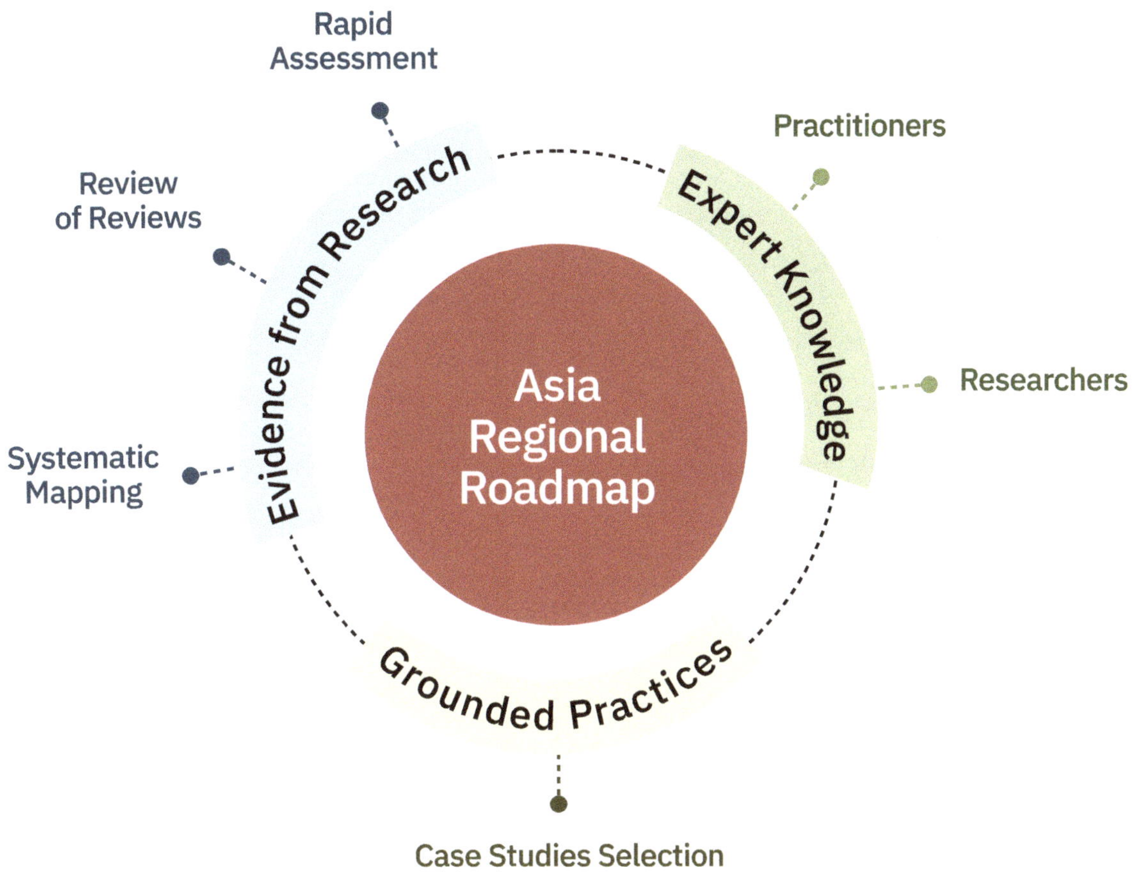

Figure 3: Integrative Approach to the NATURA Asia Roadmap for Urban Nature-based Solutions.

2.2 Methods to Assess the State of Research

To assess trends and gaps in current urban NbS research, the team applied a three-part methodology designed to capture both broad patterns and detailed insights across the field. These methods formed part of a coordinated approach used in regional assessments for the NATURA Global Roadmap, enabling comparative analysis across regions and the global literature. This approach synthesized publication trends, thematic focus, and geographic distribution in the peer-reviewed literature. Grey literature was also included selectively to fill gaps where peer-reviewed material was scarce. Literature reviews included a rapid assessment, a review of reviews and a systematic mapping (see Appendix A for details).

Rapid Assessment

The team conducted a broad, high-level scan of English-language academic literature on urban NbS and related concepts (e.g., ecosystem services, green infrastructure, etc.), mapping when and where NbS appeared in comparison to similar terms and highlighting regional hotspots and publication trends over time. A systematic search was carried out in Scopus and bibliometric analyses were performed to examine the evolution and distribution of the literature.

Review of Reviews

A targeted review was undertaken to synthesize trends, knowledge gaps, and recurring challenges as reported in existing review articles on urban NbS. Using defined search queries, review articles were identified and screened for relevance to urban contexts, NbS-related concepts, and systematic or scoping methodologies. Data extraction was guided by a structured matrix, adapted from prior work, to capture key information on NbS concepts, review types, thematic and geographic trends, and gaps across ecological, social, and governance dimensions.

Systematic Mapping

This most granular phase focused on primary studies explicitly engaging with the term "nature-based solutions" in urban contexts. Only empirical, place-based articles were retained following a rigorous screening process, resulting in a curated dataset. Geographical, bibliometric, and thematic trends were analyzed using RStudio and VOSViewer, with content analysis based on a validated text-mining approach to assess how societal challenges and ecosystem types are addressed across regions.

2.3 Methods to Assess the State of Practice

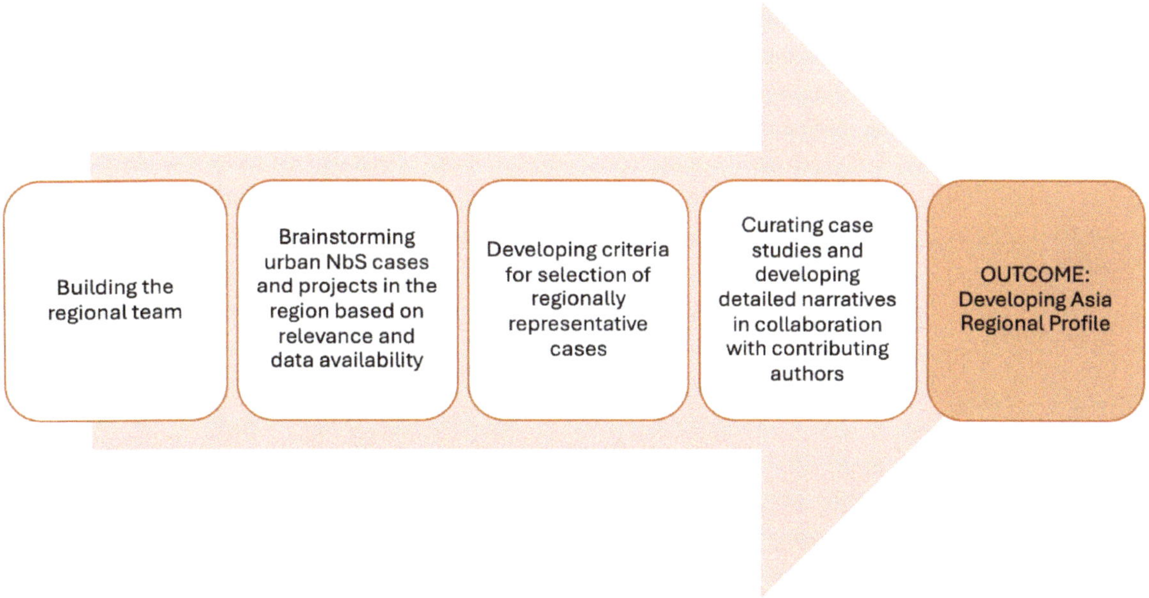

Figure 4: Steps of the Case Study Selection Methodology.

To develop the regional profile and case study narratives, the first step involved building a regional team (see Figure 4). This team included a core group of regional leads and fellows, along with contributing authors responsible for developing case studies, reviewing report sections, and offering feedback on the overall draft.

This process ran concurrently with brainstorming potential urban NbS cases and projects to include in this study. The preliminary criteria for identifying these case studies included:

- relevance to cities, either directly or indirectly, through the provision of beneficial services to urban areas, and
- the availability of sufficient, traceable data to adequately report and document the case narratives.

The initial list of case studies was developed through a combination of conversations with thematic experts, the authors' own knowledge of regional projects, and reviews of existing literature reviews and databases, such as the Urban Nature Atlas, the OECD NbS case study forum, the CDKN - Nature-based Solutions for urban climate resilience in South Asia, and the Oxford Nature-based Solutions Initiative. While the databases provided useful overviews, potential institutional bias was recognized in their selection. To address this, a snowball approach was used, engaging regional experts and reviewing additional literature to identify more projects and ensure a diverse, regionally representative set of cases.

Targeted online search was conducted using keywords such as "NbS", "Ecosystem based approaches", "Blue-green infrastructure" "Green Infrastructure", "riverine and marine restoration", "lake conservation", "coastal management", and "sponge cities" to identify additional cases. Ultimately, this process yielded over 60 NbS case studies of varying types and scales, implemented by diverse actors across more than 35 cities in 18 countries.

To shortlist case studies, a systematic procedure was established for selecting relevant and representative case studies, guided by the following selection criteria:

- **Types of ecosystems:** This criterion was used to classify the projects in terms of their ecosystems, e.g., differentiating terrestrial from freshwater and coastal projects.

- **Geographic diversity:** This criterion ensured geographic diversity by providing adequate representation of cases from various cities and countries, in addition to cities of different sizes.

- **Stakeholder diversity:** This criterion considered the range of actors engaged in the uptake and implementation of NbS practices. Initially, national governments, NGOs, industries, multilateral banks, state-sponsored agencies, and community-based organizations were identified as key actors.

- **Scale of interventions:** This criterion ensured cases spanned from individual or building-level interventions to city-wide initiatives at the scale of large metropolitan cities.

- **Objectives and outcomes of NbS:** This criterion assessed the objectives and outcomes of implemented NbS solutions, which encompassed a spectrum of goals such as climate change adaptation, climate mitigation, building urban resilience, strengthening food security, biodiversity conservation, and livelihood security.

Using these criteria as filters, case studies were qualitatively and systematically narrowed down relevant case studies within each type of ecosystem. Additional filters for country, type of stakeholder, and scale, refined the selection. This comprehensive, multi-dimensional approach allowed us to compile a set of case studies that represents the rich diversity of regional contexts and NbS practice. The process resulted in a set of 27 case studies from 16 countries that are representative of the regions and the different types of NbS implemented. Case study narratives for each of these shortlisted cases were developed in collaboration with the authors of this report.

While this process aimed to ensure inclusivity and regional diversity, certain limitations were recognized: first, the team faced language barriers and used translating software to access reports in regional languages. Second, while the team included NbS practitioners, it was led by academics, and cases led by government actors were not captured directly. As a result, the final database may reflect some geographical and thematic biases and should not be used for statistical inferences. Instead, it serves as a curated collection of examples that illustrate trends in urban NbS across Asia. The full list of case studies is provided in Appendix.

2.4 Team and Positionality

This report is the result of a collaborative study led by a core team of four researchers who consulted with 13 experts from 11 countries, each bringing a wealth of experience in urban NbS and related fields across Asia. The team brings diverse perspectives from their disciplinary backgrounds — spanning ecology, urban planning, climate change adaptation, community engagement, sociology, civil engineering, hydrology, geography, and other fields.

Drawing on a synthesis of disciplinary knowledge and practical insights, the report reflects contributions from both academic and practitioner communities. We recognize, however, that no single study can fully represent the rich complexity of ecosystems, cultures, and governance frameworks across Asia. While our team's experiences, personal identities, geographic locations including ethnic and cultural backgrounds, helped shape our interpretations, we acknowledge the blind spots that arise from the limits of our representation. We are especially mindful of the need to amplify grassroots, local, and Indigenous voices that remain underrepresented in current NbS-related research and practice. Throughout the report's development, we have actively sought to be open and inclusive to diverse perspectives from various contexts. Our goal is to present findings informed not only by our collective expertise but also sensitivity and commitment to the broader cultural, geographic, and socio-political equity and diversity within Asia. We facilitated this through regular online meetings within our team, with other regional teams, and with the global coordination group. Our in-person meeting in Miami in 2023 provided a valuable opportunity for all GRM members to review each other's work, compare interpretations of the findings, and learn across teams (see image below). Addressing gaps is an ongoing challenge, but also a critical opportunity for advancing more inclusive and equitable regional assessments.

Asia team authors in Miami during one of NATURA Global Roadmap for Urban Nature-based Solutions cross-regional workshops. Urban Systems Lab, 2023.

3 State of Urban Nature-based Solutions Research and Practice in Asia

Urban nature in the Zhejiang Province, China. Yuan Wang.

3.1 Status and Trends

3.1.1 Adoption of Nature-based Solutions and Related Concepts in the Region

KEY FINDING 1:

Urban NbS adoption in Asia is expanding rapidly, yet progress remains uneven across subregions. East, South, and parts of Southeast Asia show the strongest uptake of urban NbS experimentation, while Central Asia exhibits limited adoption. The proliferation of diverse terms (e.g., sponge cities, *satoyama*, water sensitive design, blue-green infrastructure) reflects both innovation and fragmentation across the region.

Adoption and Diffusion

NbS, as a concept, has gained significant popularity throughout Asia over the last decade, especially in rapidly growing cities. The review of reviews indicates a marked rise in peer-reviewed publications on urban NbS in most Asian countries since 2019 (see Figure 5).

The term has also become part of discussions and debates about climate adaptation and urban renewal, supported by international funders such as the World Bank and the Rockefeller foundation (World Bank, 2023b; ACCCRN, 2014). Demonstration projects in countries—including China, India, Singapore, South Korea and Japan—have further promoted the adoption of NbS in the region (Luo et al., 2023). While there are many urban NbS examples in East and South Asia, other parts of the region, such as Central Asia, have a limited number of such initiatives.

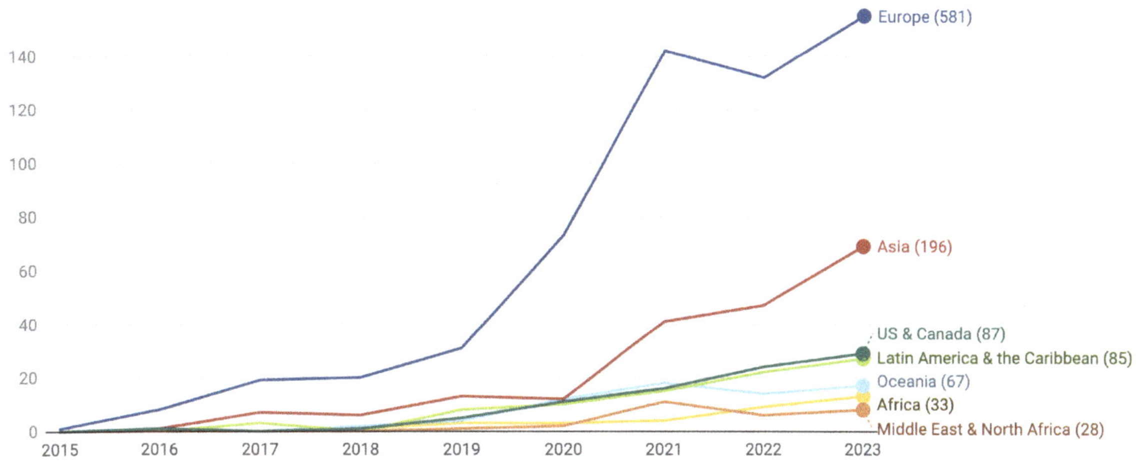

Figure 5: Regional Distribution of Empirical Academic Literature on Urban Nature-based Solutions (2015–2023)

An emerging group of urban centers in countries like Bhutan, Vietnam, Bangladesh, Malaysia, Thailand and Indonesia is actively promoting the use of NbS to tackle urban challenges. Examples include investments in new "forest parks" in Thailand, such as the Benchakitti Forest Park[1] which transformed brownfield sites into quality public spaces (Likitswat and Sung, 2025). In Indonesia, small-scale "biopores"—water infiltration systems—are being implemented to improve stormwater management (Hamel and Tan, 2022). Other countries including Indonesia, Singapore, and Malaysia, are advancing larger city-scale projects (Yuanita et al., 2025). On similar lines, the Bhutanese government has launched an 8.9 million USD project aimed at enhancing climate resilience in its water sector through NbS (United Nations Development Programme, 2023). In other countries, like Japan, Satoyama practices demonstrate how NbS can integrate traditional management strategies used historically in rural areas and strengthen meaningful human-nature connections in peri-urban contexts (Mabon, 2019).

Benchakitti Forest Park in Bangkok, Thailand. Erich Wolff, 2023.

Satoyama landscape in the Minamisanriku-Town, Miyagi prefecture, Japan. Yasuo Takahashi.

[1] See case AS_01

In contrast, a consistent adoption of NbS has not yet been observed in Central Asia. Several factors may explain the slower uptake, including regional climate conditions that limit the feasibility of certain types of solutions (Adambekov et al., 2016; Zhang et al., 2021). In countries such as Kyrgyzstan, Tajikistan, and Turkmenistan, urban NbS are not yet well documented, as development policies often prioritize restoring rural landscapes and protecting traditional livelihoods (World Bank, 2023b). Additionally, some documented cases exist only in local languages, making details about the terminology, implementation, governance, and monitoring practices largely inaccessible to international audiences.

Terminology and Themes

The current terminology used to describe urban NbS in Asia has emerged as an extension of other commonly used terms, including blue-green infrastructure, sponge cities, water sensitive urban design, *satoyama*, and ecosystem-based approaches. These ideas draw from local experiences, while also being influenced by North American, Australian and European influences (Pauleit et al., 2017).

While the concepts of blue-green infrastructure and water-sensitive urban design primarily originate from engineering and urban planning literature, ecosystem-based approaches draw from ecology and emphasize the preservation and restoration of ecosystems for the services they provide (Cohen-Shacham et al., 2016). Other traditional and emerging regional concepts such as *satoyama* in Japan and sponge cities in China provide insights into how the broader field of NbS has been adopted or re-interpreted based on local needs and specificities in cities across Asia. Overall, the field of NbS in Asia builds on this diverse foundation by integrating biodiversity and ecosystem services into urban planning and design, aiming to enhance sustainability and resilience within urban environments (Nesshöver et al., 2017).

The dominance of NbS for water management in the region is reflected in the language used to describe NbS projects. Government agencies, urban planners, architects, and environmental organizations in the Asia-Pacific region often use terms like water sensitive urban design for projects involving wetland systems and rain gardens (Irvine, Ho and Chua, 2021). Similarly, blue-green infrastructure,

West Lake Cultural Landscape of Hangzhou, Zhejiang Province (China). Yuan Wang.

among other country-specific terms, describe approaches focused on freshwater systems that often contain components of NbS (Hamel and Tan, 2022). At the national level, other locally adapted terms have also emerged. For example, in Thailand, “monkey cheeks” refers to water-retention areas designed to temporarily store floodwater, much like monkeys store food in their cheeks (Hamel and Tan, 2022). Despite their differences, all these concepts share a focus on harnessing natural ecosystem functions to address environmental challenges and promote sustainability (see Figure 6).

In countries like Japan, Indonesia and Myanmar, NbS are closely connected with Indigenous and traditional practices. Examples include protecting floating gardens in Myanmar[2] and the maintenance of terraced

[2] *See case AS_10*

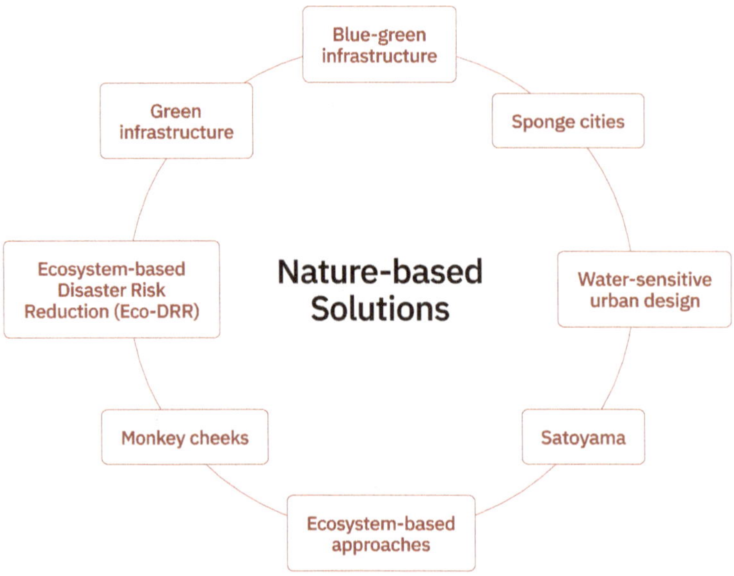

Figure 6: Illustration of Various Nature-based Solutions (NbS) Concepts (but not limited to) Used in Asian Contexts.

farming practices in Indonesia, both of which help slow runoff and reduce the risk of floods in urban catchments. Similarly in Japan, urban NbS approaches are informed by concepts that are deeply rooted in culture and locally developed practices, such as Miyawaki forests and satoyama landscapes in urban areas.[3] The satoyama concept refers to semi-natural peri-urban ecosystems that support biodiversity while also benefiting society (Iwachido et al., 2020). Such traditional and Indigenous land management practices highlight strong cultural ties to the land and create opportunities for meaningful community participation in implementing NbS (Sato, Minamoto and Ushimaru, 2023).

Rice paddies produce food and reduce the risk of floods in the periphery of Chiang Mai (Thailand). Erich Wolff, 2024.

[1] *See case AS_12*

In peer-reviewed research, however, the most commonly used NbS-related terms remain ecosystem services, green infrastructure, and urban greening (Figure 7). This illustrates a misalignment: while academic work tends to privilege globally standardized NbS-related frameworks, on-the-ground practice across Asia uses different terminologies rooted in social-ecological conditions and cultural meanings of nature.

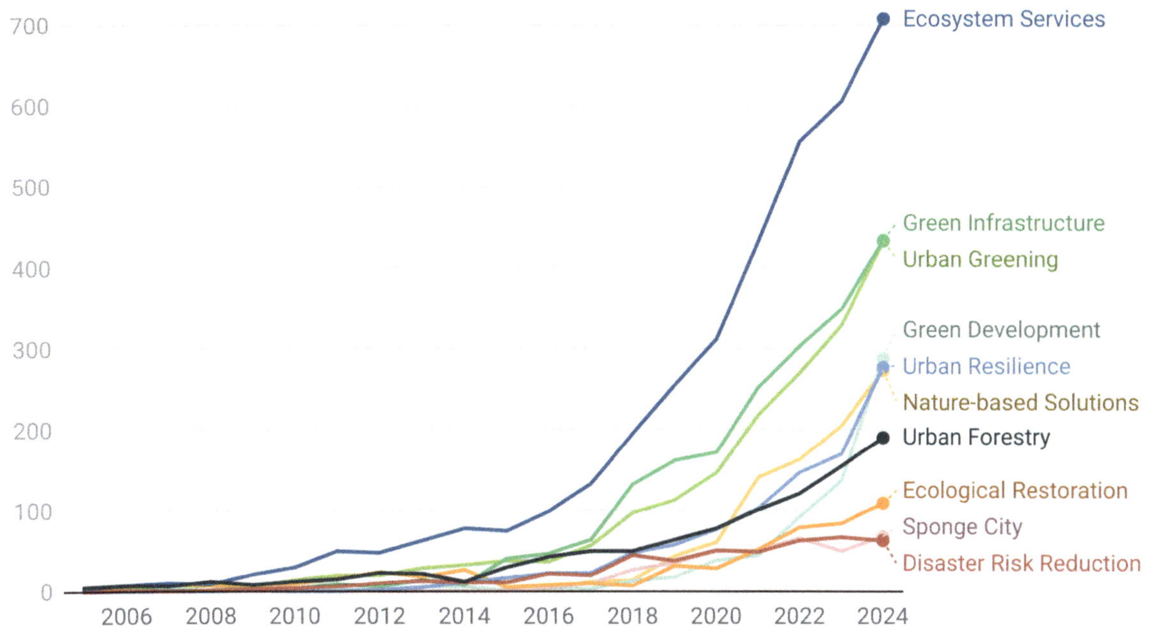

Figure 7: Evolution of the Top 10 NbS-Related Concepts in Urban Asia Research (2005-2024)
Note: Only papers that include a selected concept that appear in the tile, abstract, or author keywords of an article appear here.

3.1.2 | Types of Urban Nature-based Solutions Projects

KEY FINDING 2:

Urban NbS take diverse forms across urban and peri-urban landscapes. Parks, urban forests and green roofs appear to be most common, while river rehabilitation, coastal restoration and hybrid approaches are gaining prominence. They range from large flagship projects to community-driven initiatives operating across multiple scales, ecosystems and institutional contexts.

A Rich Mosaic of Urban and Peri-urban Nature-based Solutions

Asia's NbS reflect a broad spectrum of approaches extending from dense urban city centers to peri-urban edges, and coastal and riverine systems, responding to rapid urbanization and cities in highly diverse climates These varied conditions have produced projects that differ widely in scale, ecosystem type, and purpose, underscoring that urban systems cannot be solely within administrative boundaries (Figure 8). The review of empirical studies on urban NbS in Asia indicates the prominence of research on green ecosystem-based measures (e.g. parks, urban forests, green roofs, and vegetated corridors) followed by blue ecosystems (e.g. freshwater and marine/coastal ecosystems), blue-green ecosystems (e.g. wetlands, mangroves, hybrid stormwater management systems) and brown ecosystems (e.g., brownfields) which are less studied (Figure 9).

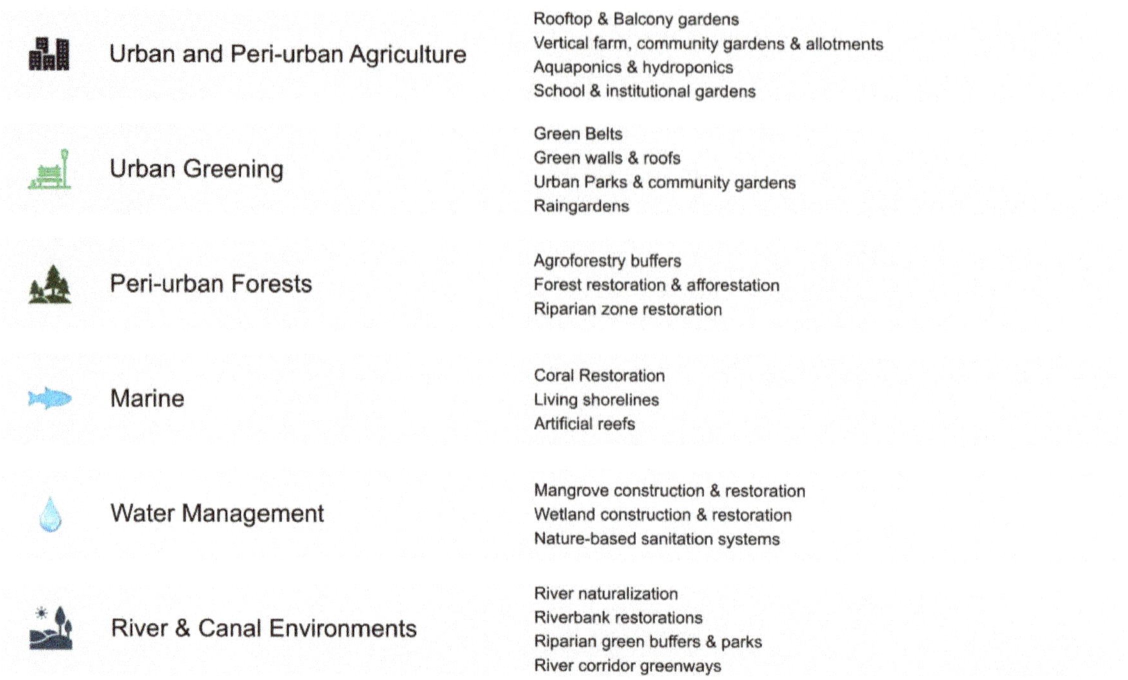

Figure 8: Common Types of Urban Nature-based Solutions in Asia.

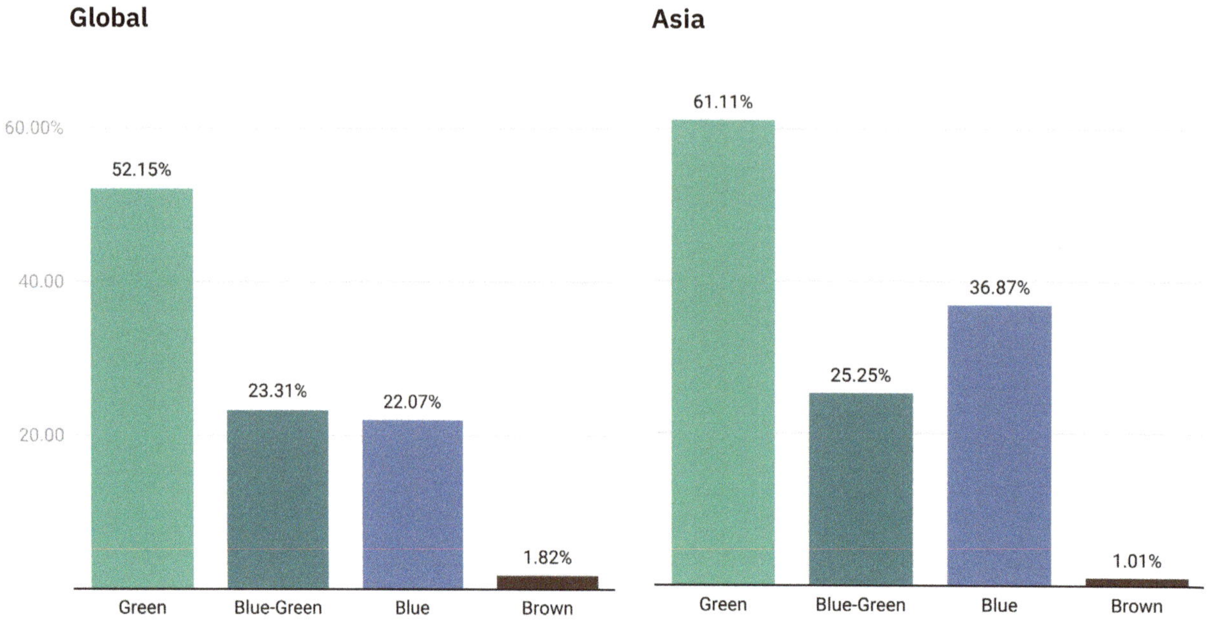

Figure 9: Proportion of Ecosystem Types in Empirical Urban NbS Studies: Global vs. Asia.

Ecological Diversity and Regional Character

Urban NbS in Asia draw on a wide range of ecosystems and reflect strong regional differentiation:

- Green and terrestrial systems dominate, including parks, forests, and green roofs.
- Coastal and riverine measures feature prominently in deltas and island nations, such as mangrove restoration, wetland rehabilitation, and coral reef protection.
- Central Asian cities exhibit limited but emerging examples, constrained by climatic and hydrological conditions.

Large-scale projects often focus on urban forests, green corridors, and coral restoration initiatives whose benefits extend beyond city limits. Such projects highlight the close interconnections between urban centres and surrounding landscapes—often described as desakota regions (McGee, 1991; Sui and Zeng, 2001)—where rural and urban uses overlap. The Korea–Mongolia Reforestation Initiative[4] exemplifies how urban NbS can connect ecosystems across borders. Similarly, projects in less populated areas—such as agriculture related NbS in Nepal, Vietnam, and Indonesia—remain relevant to urban sustainability because their outcomes (e.g., erosion control or water retention) benefit nearby cities or downstream communities.[5] Together, these examples show that NbS in Asia are not confined to city cores but are part of broader ecological networks shaped by climate, geography, and land-use dynamics.

From Flagship Redevelopments to Everyday Greening

NbS in Asia are as diverse as their local contexts of implementation, ranging from flagship urban greening projects in city centers led by multiple actors (state, municipalities, NGOs, and private developers), to informal settlement upgrading projects in urban peripheries (Wolff, Rauf and Hamel, 2023). NbS in Asia therefore operate across a wide range of scales and degrees of formality, from state-led infrastructure to community-driven initiatives:

- Flagship regeneration projects such as Seoul's Cheonggyecheon Park, Singapore's Bishan–Ang Mo Kio Park, and China's Sponge City pilots demonstrate large-scale state investment.
- Municipal and mid-scale programs promote rooftop gardens by Pune Municipal Corporation (Singh, Koduganti and Patil, 2023), green corridors and urban forests.
- Community-led actions in informal settlements implement smaller, adaptive measures such as rain gardens, urban farming, and household-level greening practices.

In East Asia, there is a strong focus on urban greening projects and stream management, often spanning building, neighborhood, and city scales. Examples include Cheonggyecheon Park in Seoul, South Korea,[6] the renowned rooftop gardens in Fukuoka Prefecture, Japan[7] and China's Gangxia 1980 Project designed to support youth mental health through green roofs[8]. China remains at the forefront of Sponge City policy, investing in city-wide, water-focused NbS such as the Chongqing Urban Regeneration Project[9] and Shijiu Yang Wetland Park.[10]

In Southeast Asia, partly due to the tropical climate and geographic position, the most common approaches included green roofs to manage local runoffs, peri-urban practices that retain soil and improve downstream water quality, and river renaturalization projects such

[4] *See case AS_26*
[5] *For example, case AS_04*
[6] *See case AS_27*
[7] *See case AS_16*
[8] *See case AS_08*
[9] *See case AS_03*
[10] *See case AS_09*

as Singapore's Bishan-Ang Mo Kio Park[11] and Thailand's Puey Ungphakorn Centenary Hall Green Roof.[12]

In Central and South Asia, a range of local or city level urban greening initiatives (green spaces, parks), urban agriculture (edible and recreational gardens), and wetland conservation projects are more prominent (Rao et al., 2022), as exemplified by case studies like Re-greening Tashkent in Uzbekistan, Ragi Project in India, and Oyster Reef Barrier in Bangladesh[13].

Local implementation of NbS is also shaped by the international policy agendas on climate, water, and infrastructure (Lakshmisha, Nazar and Nagendra, 2024; Hamel and Tan, 2022) and unlocking innovative financial mechanisms is seen as a priority by stakeholders such as multilateral banks (Asian Development Bank, 2024). However, implementation faces several challenges. For instance, existing fragmented governance structures often pose barriers for mainstreaming strategies, which is aggravated by shifting priorities and changes in the political landscape that may disrupt program continuity. These governance changes can slow down the implementation of NbS and climate adaptation strategies, especially in rapidly growing cities in South and Southeast Asia, characterized by elevated rates of informal urbanization.

Typical challenges of informal urbanization processes—such as insecure land tenure, limited space and scarce resources—can delay formal adoption of NbS, even though these solutions could greatly benefit vulnerable populations (Greenwalt et al., 2023; Wolff, Rauf and Hamel, 2023). Despite these barriers, many informal settlements in South and Southeast Asia actively implement NbS rooted in traditional knowledge and practices (Chatterjee, 2010; Dangol and Day, 2017). These community-led approaches to urban greening, water conservation, and biodiversity management often diverge from formal municipal urban planning strategies. Leveraging existing practices and locally available material and plants, communities develop NbS that improve access to food, water and mitigate climatic hazards (Greenwalt et al., 2023; Patil et al., Under review).

Green cover in Mandalay, Myanmar's second largest city. Chandni Singh, 2017.

[11] *See case AS_02*

[12] *See case AS_14*

[13] *See case AS_19; AS_17; AS_25*

Co-benefits within Diverse Social, Environmental, Ecological Contexts

Urban NbS in Asia are being increasingly used to tackle multiple urban challenges, particularly those related to climate change and the Sustainable Development Goals (SDGs), from flood and heat risks to poverty reduction. Most NbS projects reviewed aimed to enhance climate resilience, including initiatives like urban forests, green roofs, and biodiversity corridors in Seoul, Bengaluru and Jakarta. They improve water security through mangroves and wetland conservation in Lahore, Pakistan, Kolkata, India, and Putrajaya, support biodiversity through urban parks in Bangkok, Thailand, New Delhi, India and contribute to infrastructure development (e.g., Hatirjheel Integrated Development Project in Dhaka) (Urban Nature Atlas, 2024).

Societal priorities vary by context in different regions of Asia. Cities with limited space like Singapore, focus on addressing floods and regulating temperature through green roofs, vertical gardens, and water management systems. Countries like China, South Korea, and India rely on NbS to address air pollution, urban heat islands, and climate extremes through green corridors and tree planting. Meanwhile, Japan emphasizes hazard resilience to floods and tsunamis integrating coastal green infrastructure, protective green belts, and disaster-resilient urban design into urban centers.

Beyond their direct environmental benefits, NbS in Asia also offer significant social, cultural and spiritual co-benefits rooted in relational and intrinsic views of nature. These co-benefits are often derived from the personal and social bonds communities have with nature and underpin knowledge systems that acknowledge the right of flora and fauna to exist, independent of their services to society (Chan et al., 2016; Wolff et al., 2025). These relationships established between nature and people are evident in urban farming projects (see Section 3.3), grassroots initiatives in informal settlements (Rauf, Wolff, et al., 2024; Singh, Koduganti and Patil, 2025) and the development of urban green spaces and nature reserves specifically designed for biodiversity protection or restoration.14

In cities such as Ho Chi Minh City (Vietnam), cultural and spiritual connections with nature have been actively promoted. Erich Wolff, 2025.

The implementation of some NbS projects can unequally distribute the benefits and costs, leading to unintended negative outcomes. For example, some NbS projects may primarily deliver benefits to affluent neighborhoods, limiting access for lower-income residents, as observed in several climate adaptation initiatives in India, Indonesia and the Philippines (Anguelovski et al., 2016). Additionally, projects marketed as urban revitalization such as waterfront redevelopments in Thailand (Chao Phraya Promenade river) and India (Sabarmati river) can reproduce gentrification and green washing practices, attracting wealthier residents and businesses while displacing existing communities and worsening urban inequalities (Ashraf, 2020; Shaikh and Hamel, 2023). Furthermore, the integration of NbS within urban planning and smart city initiatives, like Tokyo's Smart City endeavors with urban farming apps, might disproportionately benefit affluent neighborhoods that have greater access to resources and technology, potentially excluding lower-income areas from participating in and benefiting from urban NbS.

[14] *See case AS_16*

3.1.3 | Actors and Partnerships

KEY FINDING 3:

Urban NbS in Asia are driven by layered coalitions. Municipalities often anchor projects; national agencies and multilaterals add frameworks and funding. NGOs and private firms contribute expertise. Community roles vary widely from limited consultation to co-management. Scaling remains constrained by coordination gaps and short-term funding.

A broad range of actors are involved in the adoption and implementation of NbS landscape across Asian cities. Based on the needs, capacities, and motivations of different actors, these actors can be broadly categorized into four types: national governments and international agencies; local governments and other government-related departments, NGOs, and private businesses; and local community and citizen groups (Figure 10).

Among these, local governments often take the lead, initiating projects and providing essential institutional support, land, and financing. For example, Seoul's municipal government spearheaded the restoration of Cheonggyecheon stream, while in Nepal, government agencies have supported forest conservation through dedicated funding and institutional backing. Some municipal governments also tend to allocate portions of their annual budgets to develop urban parks and green infrastructure, such as ongoing renovations in green spaces in Metro Manila (the Philippines). In addition, local or national governments may offer tax incentives or subsidies to encourage the adoption of green roofs or the construction of energy-efficient buildings, as demonstrated by the Pune Municipal Corporation (India) (The Green Bein, 2025). Beyond these direct benefits, local governments gain from NbS by enhancing their city's image, boosting municipal revenue, and fostering partnerships, as seen in urban farming initiatives in Pune.

National governments and international organizations, such as the International Union for Conservation of Nature (IUCN) and bilateral agencies and multilateral institutions, also play a vital role. They provide financial incentives, including international aids and grants, to accelerate the transition towards NbS. These actors often offer frameworks for demonstration projects and facilitate knowledge exchange. Notable examples include the East Kolkata Wetlands in India[15] and various NbS projects in China. The integration

Figure 10: Actors Involved in NbS Uptake and Implementation.

[15] *See case AS_22*

of government efforts is particularly evident in smaller countries like Singapore, where coordinated incentives are offered across multiple levels and agencies to promote NbS.

Several actors including IUCN, Deutsche Gesellschaft für Internationale Zusammenarbeit (GIZ), the Asian Development Bank (ADB), and research universities have been regularly organizing webinars, online dialogues, and knowledge-exchange sessions in collaboration with NGOs and cities (see posters below). These events provide updates on new developments in NbS, share case studies, and encourage collaboration among diverse stakeholders.

NGOs and businesses also make crucial contributions to implementing NbS in the region. They often bring essential contextual knowledge and management expertise, helping tailor solutions to local needs and creating the conditions necessary for implementing pilot projects that can later be scaled up. Funders such as the Rockefeller Foundation (ACCCRN, 2014) and companies like Urchinomics[16] can significantly influence the design and priorities of NbS projects, especially in niche sectors where NbS are not yet mainstream such as informal settlement upgrading and blue carbon credit markets.

Citizen groups, local communities, and residents are typically the most diverse and primary beneficiaries of NbS. Their roles, however, can vary widely across projects, such as the Bengaluru Climate Action Plan (BCAP) which engages with citizen groups, schools and colleges, and civil society in an informal institution called 'Friends of BCAP' ; and

Poster for the 'Nature and Climate Nexus Webinar Series: Promoting Nature-based Solutions for Sustainable Infrastructure in Asia and the Pacific', hosted by ADB. ADB, 2022

Poster of the 4th webinar under the "Sustainable Urban Development – Smart Cities -II" series organized by National Institute for Urban Affairs (NIUA), India. India Forum For Nature Based Solutions, 2023.

[16] *See case AS_07*

community forest management initiatives in Nepal and Pakistan, which leverage community leadership and decision-making though bottom-up and hybrid models.

Overall, urban NbS projects—including green infrastructure (e.g. green buildings), sustainable water management systems, and mangrove and coral restoration for hazard protection—are expanding across Asia and the broader Global South (Castaldo et al., 2025). Despite this growth, challenges remain in assessing their effectiveness and in measuring their social and economic benefits for adaptation (UNEP, 2023). It is, therefore, critical to empower and train key stakeholders—including government officials and private sector actors—on the potential to co-develop NbS in order to meet climate and biodiversity goals. Raising awareness and sharing knowledge about NbS through partnerships between different stakeholders and knowledge brokers, such as universities, is especially important, given Asia's political diversity, where funding often depends on political cycles, international cooperation, and willingness of external investors.

3.1.4 | Mechanisms for Inclusion and Participation

KEY FINDING 4:

Use of diverse participation mechanisms for NbS is increasing, yet equity and depth of involvement vary widely and often remain symbolic. Inclusive initiatives that integrate local knowledge, including gender perspectives, contrast with state-led projects that risk displacement.

Cities in Asia use a variety of mechanisms to promote inclusion and participation of diverse stakeholders in the planning and implementation of NbS. The nature and extent of participation, however, varies widely depending on local context, governance structures, and community involvement (Erbacher and Wilkens, 2024). In general, NbS implementation involves a range of stakeholders, including government authorities, local communities, NGOs, researchers, and other relevant actors. However, not all participation processes are equitable. For example, the development of Chulalongkorn Centenary Park in Thailand—a major urban green space—led to the clearance of former shophouses (a mixed building-use with shops on the ground floor with residence above) in the surrounding area and their replacement with shopping centers and upscale residential buildings. While the park is considered a landmark project in the region, this process has been denounced for contributing to gentrification and the displacement of vulnerable residents (Pottinger-Glass, 2022). Such cases highlight a broader regional challenge: the need to ensure NbS are implemented in ways that are socially just and appropriate to local contexts (Nassary et al., 2022).

It is [...] critical to empower and train key stakeholders — including government officials and private sector actors — on the potential to co-develop NbS in order to meet climate and biodiversity goals.

Gentrification has been observed near the Chulalongkorn University Centenary Park in Bangkok (Thailand). Erich Wolff, 2023.

In Dhaka, Bangladesh, the 'Urban Partnerships for Poverty Reduction' project exemplifies gender-inclusive planning by ensuring that the needs and perspectives of all genders are meaningfully considered in NbS implementation. Similarly, pilot projects such as the RISE program in Indonesia (see section 3.3) offer lessons on how inclusivity and participation can be integrated into NbS initiatives, primarily in informal urban contexts (Moschonas et al., 2025). These cases also underscore the value of fostering partnerships, while recognizing the diverse contributions and perspectives of all stakeholders in NbS planning and delivery.

While participatory approaches are gaining currency in the planning, design and implementation of NbS, some projects continue to operate within centralized governance models that limit meaningful community involvement. In certain cases, community participation has been described as 'not decisive,' with residents being 'simply informed' about NbS initiatives without having real inputs into decisions (Qiao et al., 2019). Examples across Asia show that, while strong state-led coordination and centralized leadership can help accelerate the implementation of NbS, it can also reduce opportunities for community engagement (Qiao et al., 2019; Tan and Low, 2019). This can pose a significant challenge for mainstreaming NbS in Asia, as projects that do not align with local priorities or secure community buy-in, risk being rejected, or fail to deliver lasting benefits (Mabon, 2019).

3.1.5 | Financing Mechanisms

KEY FINDING 5:

Financing for urban NbS in Asia combines grants, loans, PPPs, and emerging market instruments such as green-blue bonds and carbon and biodiversity credits, yet access varies by governance context, project scale and investor expectations. Large projects depend on multilateral funds; while smaller community or municipal projects depend on inconsistent local budgets. Thus, monetizing adaptation and public-good benefits remains a core barrier to the private sector.

Financing for urban NbS exhibits significant variability, often shaped by regional economic conditions, governance structure, and urbanization patterns. Figure 11 illustrates the wide range of investment instruments used to channel financial resources into NbS projects.

These instruments could be grant-based (such as direct grants, or redeemable grants) debt-based (private loans, green bonds, or social bonds); equity-based investments (both private and public) and derivative mechanisms (like carbon credits). These investments can operate and be traded in both public and private markets. Since many of the instruments are often complex and non-standardized, significant variations can exist both between and within categories of instruments. Their suitability depends on several factors including characteristics of investors (such as investment size, duration, return expectations, etc.) and the nature of the entity seeking capital (government, municipality, small/large business, NGO, etc.). Additionally, the type of underlying investment—whether corporate, project-based, revenue-generating, or focused on reforestation or ecosystem restoration—is a

Public Finance

- Urban Greening Park, Tokyo Japan
- Benjakitti Park in Bangkok, Thailand
- Emerald Cities, Uzbekistan

Private and public partnerships

- Sponge City Program in Shanghai, China
- Seoul Sky Garden, South Korea
- India Forum on Nature-based Solutions which is funded by Caterpillar Foundation, the Department of Environment

Grant and donor funding

- Green public procurement initiative, Bhutan
- Rockefeller Foundation's Asian Cities Climate Change Resilient Network

Development banks and financial institutes (DFI)

- Green building initiatives, Pakistan
- Community gardens financed by the Thai Health Promotion Board, Thailand

Community Financing

- Coral conservation in Colombo, Sri Lanka

Figure 11: Common NbS Finance Models in Asia.

pivotal consideration in selecting appropriate financing mechanisms.

In South and Southeast Asia, international funding and national government investments have demonstrated the potential of cross-border collaboration for NbS. Large-scale projects in the region often rely on organizations with the capacity to mobilize and access significant resources, such as development aid funds from the Rockefeller Foundation (ACCCRN, 2014) or financing from multilateral banks (Matthews and Dela Cruz, 2022). Additionally, in Mumbai (India), partnerships with private developers, and corporations have been established to develop urban parks and green spaces (Thakur Nanda, 2025). Through Public-Private Partnerships (PPPs), private entities contribute both funding and expertise, while the government ensures public access and oversees sustainable management.

While these projects offer important insights into scaling up NbS in Asia, more research is necessary to assess how well they align with the needs and priorities of the most vulnerable communities. Smaller-scale NbS initiatives by local governments, such as the terrace garden in the Pune Municipal Corporation, or those led by civil society organizations and grassroots movements, would benefit from innovative policies to support mainstreaming (Rauf, Wolff, et al., 2024; Singh, Koduganti and Patil, 2023). Scaling up and scaling out these projects is key, as community gardens, tree-planting programs, and similar efforts often emerge in response to local needs and residents' aspirations (Kharisma and Kaswanto, 2021). Many of these projects have reported promising results, in terms of local engagement and positive outcomes, for communities and ecosystems (Urban Poor Development Fund and Asian Coalition for Housing Rights, 2010). However, gaps remain in understanding how such efforts can be systematically scaled and mainstreamed. As a result, there is a pressing need to develop innovative governance models that reflect the unique political, social and economic drivers shaping the landscape of NbS in Asia (Musadat, 2019).

Green roof in a private commercial development in Singapore. Erich Wolff, 2025.

Solar panels installed with their share being used for growing edible plants at ward-level office of the Pune Municipal Corporation, Pune, India. Maitreyi Koduganti, 2025.

It is also essential to recognize that NbS vary widely in their financial characteristics, and not all solutions offer equal investment potential. Currently, most nature-related investments are directed toward gray infrastructure within established economic sectors that provide well-understood and appealing commercial returns (such as forestry, agriculture, tourism.) One of the main challenges in financing adaptation-focused NbS lies in the difficulty of monetizing their benefits, which are typically public goods that are 'non-excludable' and accessible to all (Khan and Munira, 2021). This differs from the more straightforward monetization of greenhouse gas emissions reduction, which more readily supports projects, offering traditional commercial revenues or carbon credits returns. While non-commercial benefits such as social and biodiversity gains can attract demand and even command a premium from some investors, overall financial flows tend to favor investments with higher returns relative to risk (Van Raalte and Ranger, 2023).

3.1.6 | Policies, Legal Frameworks, and Urban Strategies

KEY FINDING 6:

NbS in many Asian cities are closely tied to urban planning frameworks often anchored within climate action, land-use planning, smart city, and sustainability strategies. Examples like Singapore's City in Nature, Jakarta's Urban Greening Program, India's Nagar Van (Urban Forest) Scheme, and China's Sponge City initiative show strong national alignment. Yet, integration remains fragmented, with limited legal instruments and weak monitoring systems.

NbS projects implemented in urban areas across Asia are often embedded within broader policies, regulations, and urban strategies designed to promote environmental sustainability, bolstering climate resilience, and fostering urban development. Many cities have adopted comprehensive frameworks that influence or directly incorporate NbS initiatives. For example, cities such as Mumbai, Chennai, and Bengaluru in India, Jakarta in Indonesia, and Seoul in South Korea, have integrated NbS projects into their climate action plans to enhance climate resilience (see box 3). Similarly, cities pursuing 'Smart City' policies

— including Hangzhou (China), Bangkok (Thailand), Indore (India), Nusantara (the new capital of Indonesia) and Tokyo (Japan) — have also incorporated NbS elements in their urban development strategies (Irvine et al., 2023). Table 1 highlights key urban NbS policies, governance structures, financing mechanisms, and challenges in selected Asian countries.

Table 1: Key NbS Policies, Governance Structures, Financing Mechanisms, and Challenges in Asian Countries

Country	Key NbS policy or strategy	Policy focus	Governance structure	Financing mechanism	Challenges and Gaps
China	Sponge City Initiative (2015)	Urban flood management via low-impact development, retention & reuse	Central guidance by Ministry of Housing and Urban–Rural Development (MOHURD), Ministry of Finance (MOF), Ministry of Water Resources (MWR); municipal pilot implementation	Central subsidies for pilots; local budgets; PPPs; some projects complemented by multilateral loans (ADB)	High retrofitting/O&M costs; uneven local capacity; monitoring/performance evaluation issues.
India	Smart Cities Mission	Urban renewal incl. liveability, water, mobility, public spaces (blue-green integration varies by city)	Ministry of Housing and Urban Affairs (MOHUA) oversight; city-level Special Purpose Vehicles with State/Urban Local Bodies (ULB) participation	Government of India grant of 5B INR per city (matched by state/ULB); PPPs; convergence with schemes	Delays, coordination gaps, capacity constraints in Special Purpose Vehicles (SPVs) and ULBs.
	East Kolkata Wetlands (EKW) Programme	Wetland conservation & wastewater reuse ecosystem; urban flood buffer	East Kolkata Wetlands Management Authority (Govt. of West Bengal); statutory protection	Primarily state budget & regulatory instruments for protection/management	Encroachment,
	Nagar Van (Urban Forest) Scheme	Urban afforestation(i.e., 400 urban vans and 200 Nagar vatikas across the nation during 2020-21 to 2024-25) ; city-level green buffers for climate resilience	Ministry of Environment, Forests and Climate Change (MoEFCC) as nodal ministry; State Forest Departments and Urban Local Bodies (ULB)	Funded primarily under the Compensatory Afforestation Fund Management and Planning Authority (CAMPA), along with municipal/CSR support for creation & upkeep	land-use pressure; enforcement & coordination challenges.
Singapore	City in Nature Plan (2020)	Integrated blue-green infrastructure, biodiversity conservation; ecological connectivity	Centralized planning with cross-agency coordination	National budgetary funding; can align with sovereign green bonds framework for eligible projects	Land scarcity, poor maintenance; species selection issues; limited monitoring
	ABC Waters Program	Water-sensitive urban design; naturalizing canals/ponds; multi-functional waterways	Singapore's National Water Agency (PUB) leads with planning guidance; adoption across agencies/private sector	Public works budgets; private development compliance & certification pathways	Land scarcity, balancing development pressures; habitat fragmentation in a dense city-state.

Country	Key NbS policy or strategy	Policy focus	Governance structure	Financing mechanism	Challenges and Gaps
Japan	Green Infrastructure (GI) Strategy (2021)	Eco-DRR, connectivity, multifunctional urban green/ blue	Ministry of Land, Infrastructure, Transport and Tourism (MLIT), Ministry of Environment (MOE); promote GI; implementation by prefectures/municipalities	National budgets; promotion of private finance via PFI/PPP frameworks	Mainstreaming GI across sectors; aligning finance with nature-positive outcomes.
	Disaster Resilient Smart Cities Plan	Disaster prevention/mitigation integrated with smart urban systems	Cabinet Office leads National Resilience; MLIT/local govts execute	Multi-year national plans (e.g., Five-Year Acceleration Plan); standard public budgets; PPP where suitable	Aging infrastructure; multi-hazard exposure; coordination across levels.
South Korea	Garden city Seoul Initiative,	Expanding parks, street trees, green roofs/walls to improve liveability & climate resilience	Seoul Metropolitan Government with district offices & civic participation	Municipal budgets; incentives/partnerships	Space constraints in dense districts; long-term maintenance.
	National Urban Forest Plan (KFS)	Urban forests & green networks for health, climate & disaster mitigation	Korea Forest Service sets policy; municipalities implement	National & local budgets; programmatic funds	Ensuring equitable distribution & maintenance of urban forests.
Indonesia	Jakarta Green Infra Policy	Urban flood control, infiltration, retention, WSUD practices	National Ministry of Public Works and Housing (PUPR) regulations set drainage master planning; DKI Jakarta agencies (BAPPEDA, Water Resources) implement	National & provincial budgets; programmatic public investment	Informal settlement constraints; O&M capacity; legacy grey infrastructure.
Vietnam	Water-sensitive Urban Design Policy - emerging in policy and pilots	Integrating WSUD in urban drainage, heat mitigation, public spaces	National strategies (Ministry of Construction (MOC), Ministry of Natural Resources & Environment (MONRE); cities pilot WSUD approaches	State public investment; alignment with national climate/urban strategies	Limited guidance & capacity; fragmented implementation
	Climate Resilient Cities Initiative ((e.g., Mekong Delta under Res. 120/NQ-CP)	Climate-resilient regional/ urban development, water management, nature-based adaptation	Government Resolution 120 (2017); Mekong Delta Coordination Council; line ministries & provinces	Public investment programs; mobilization of domestic/ODA finance through national strategies	Cross-provincial coordination; mainstreaming resilience into urban plans.

Country	Key NbS policy or strategy	Policy focus	Governance structure	Financing mechanism	Challenges and Gaps
Bangladesh	Bangladesh Delta Plan 2100	Integrated water-land management; flood risk reduction; coastal resilience; NbS	National planning with local implementation	Public Investment Program; programmatic & project funding	High climate vulnerability, funding sustainability, local participation, Implementation capacity; coordination across agencies & delta zones.
	Coastal Green Belt Project	Mangrove/ green belts for cyclone/storm-surge protection; livelihoods	Bangladesh Forest Department; social forestry/user groups	Govt. programs; project financing over time	Land tenure/ encroachment; species selection & survival; maintenance.
Thailand	Green Bangkok 2030 incl. green-roof/ greening actions	Increase urban green space; heat & flood risk reduction	Bangkok Metropolitan Administration (BMA) with partners	Municipal budgets; partnerships & incentives	Urban heat/flood risk; maintenance & limited space for greening.
	Urban Tree Planting/ Greening Program	Street trees, parks, pocket parks towards GB2030 targets	BMA departments; district offices; civic partners	Municipal budget; corporate/ community partnerships	Ensuring equitable distribution; long-term care.
Pakistan	Ten Billion Tree Tsunami (2019)	Large-scale afforestation, restoration & EbA	Federal-provincial coordination with local implementation	Federal Public Sector Development Programme (PSDP) & programme funds; provincial co-financing	Monitoring, survival rates, governance of large-scale plantings
	Green Stimulus Package	Nature-based employment & restoration during COVID recovery	Ministry of Climate Change (MoCC); provincial partners	Stimulus allocations via federal budget; program funds	Sustained financing & verification of outcomes.
Mongolia	Korea-Mongolia Green Belt Project	Desertification control, wind erosion mitigation	Bilateral cooperation (Mongolia–South Korea), national agencies	Korean government support, international funding	Harsh climate conditions, long-term maintenance, local capacity building

Regional regulations and plans aimed to achieve the SDGs, often include measures such as land-use and zoning, and water management strategies, that support green spaces, stormwater management, wetlands restoration, and green corridors. For example, Singapore's transition to its 'City in Nature' strategic vision represents a landmark urban policy that promotes vertical gardens and green roofs. By offering financial and regulatory incentives to developers, the city encourages alignment with national goals to expand urban canopy cover and address biodiversity conservation, climate resilience, and broader sustainability objectives. This approach is also reflected in initiatives like Jakarta's 'Urban Greening Program', and Shanghai's Sponge City initiatives.

3.2 Challenges to Implementation

Urban NbS in Asia face a variety of challenges that hinder their effective design, implementation, and long-term sustainability. These challenges often stem from issues of community engagement, governance, financing, technical capacity, and evaluation.

KEY FINDING 7:

Community engagement in NbS remains limited due to language barriers, insecure tenure and top-down planning. These gaps weaken trust, legitimacy, while paving the way for greenwashing which exacerbates social inequalities, particularly in contexts of urban informality.

Many NbS projects struggle with limited community engagement, due to language barriers, limited awareness, and unequal power dynamics. This is especially problematic where collective assets and strong community bonds are crucial to building resilience (Archer et al., 2020).

Building trust in NbS in Asian cities requires stakeholders to prioritize local needs, uphold environmental justice, and actively guard against green-washing. Persistent greenwashing practices can undermine the potential of NbS by reproducing existing inequalities and reinforcing power imbalances in the implementation of climate adaptation strategies (Anguelovski et al., 2016; Yarina, 2018). To prevent this, it is crucial to ensure that NbS are not misused as a means to legitimize urban development projects that encroach on vulnerable ecosystems (Shaikh and Hamel, 2023) and to recognize that greening initiatives cannot be positioned as a substitute for comprehensive and ambitious climate action (Seddon et al., 2021; Singh, Madhavan, et al., 2021).

In informal settlements, inadequate funding for public consultations and insufficient support for community-led initiatives often limited resident involvement (Erbacher and Wilkens, 2024; Anguelovski et al., 2016). A lack of secure tenure can also discourage investments in NbS. In other contexts, centralized planning approaches and limited participation mechanisms create a gap between NbS interventions and the actual needs of residents (Møller et al., 2012; Shannon and Legrand, 2007). These challenges illustrate how historically marginalized groups, such as informal settlement residents, often have limited representation in decision-making forums for climate adaptation plans. As a result, adaptation strategies may fail to reflect the diverse perspectives and needs of these communities (Yarina, 2018; Salim, Bettinger and Fisher, 2019; Wolff, Rauf and Hamel, 2023).

KEY FINDING 8:

Weak cross-sectoral governance, fragmented institutions, and short-funding cycles undermine long-term impact. Limited monitoring capacity, rapid implementation pressures, and inadequate investment in sustained community engagement, handover planning, and uneven capacity constrain efforts to promote and coordinate NbS.

There is a pressing need for stronger investments in cross-sectoral governance both at local and national levels to support NbS and climate adaptation strategies. The diversity of stakeholders, institutional fragmentation, and inadequate monitoring and evaluation capacity in Asian cities further complicate efforts to promote and scale NbS (Lechner et al., 2020).

Funding mechanisms often rely on short-term grants and crowdsourcing to support small-scale, local NbS experiments. While these can seed innovation, they can also hinder meaningful community participation by prioritizing rapid implementation over sustained engagement and co-design (Wolff, Rauf and Hamel, 2023). The funding and governance model of many projects tend to direct limited resources toward immediate construction, leaving few funds for long-term community involvement, planning for project handover, or capacity-building among local stakeholders.

KEY FINDING 9:

Many Asian cities continue to favor grey infrastructure, partly due to skepticism and limited evidence about NbS effectiveness, as well as limited technical capacity. These trust and capacity gaps reinforce reliance on conventional engineering, underscoring the need for robust evaluation mechanisms and stronger knowledge dissemination to demonstrate NbS benefits.

Despite the growing interest in urban NbS as a strategy for urban placemaking and resilience, many Asian cities continue to favor traditional gray infrastructure solutions. This preference often reflects skepticism about NbS effectiveness, inadequate technical capacity, and limited robust evidence from existing NbS projects (Prana and Susilo, 2020; Shaw et al., 2022).

Combating these barriers requires investment in robust evaluation mechanisms to demonstrate that NbS can deliver cumulative, long-term benefits that may not be as immediately visible as conventional engineering solutions (Dumitru, Frantzeskaki and Collier, 2020). Disseminating scientific knowledge and building trust among stakeholders is essential to secure the broader acceptance of NbS as a core element of climate adaptation strategies.

KEY FINDING 10:

Weak monitoring and evaluation systems limit evidence for NbS effectiveness, learning, and investment across Asia. Without robust biophysical and social indicators, projects struggle to attract sustainable funding or refine strategies. M&E is impaired by inadequate comparative learning.

Deficient evaluation and monitoring systems for NbS in Asia present persistent obstacles, particularly for securing sustainable funding. Without clear evidence for their effectiveness, NbS projects struggle to attract investors interested in promoting ecosystem services, climate change adaptation, and reducing environmental stressors in urban areas (Shaw et al., 2022; Thompson et al., 2023). Moreover, difficulties remain in comprehending the true value and limitations of NbS for addressing climate change and other global challenges, even in countries with robust governance and moderate financing capacity (Seddon et al., 2020).

Countries in the early stages of NbS implementation can benefit significantly from investing in robust monitoring and evaluation tools. These tools help practitioners learn from their experiences, refine their strategies and compare results with case studies across the region. Examples from China and Singapore demonstrate the benefits of integrating monitoring and evaluation into water management and urban greening programs (Luo et al., 2023; Linh, Ahmed and Loc, 2022; Qiao et al., 2019). Expanding such efforts within and across countries is essential not only for generating best practices but also for supporting their adaptation in local contexts. Importantly, these monitoring efforts should assess the biophysical performance of NbS, in addition to their integration into community life and their contribution to social cohesion and resilience.

Deficient evaluation and monitoring systems for NbS in Asia present persistent obstacles, particularly for securing sustainable funding. Countries in the early stages of NbS implementation can benefit significantly from investing in robust monitoring and evaluation tools. These tools help practitioners learn from their experiences, refine their strategies and compare results with case studies across the region.

Urban green parks, such as in Yangon city, Myanmar, serve as a critical green space in dense cities across Asia. Chandni Singh, 2017.

3.3 Bright Spots

This section highlights notable NbS initiatives and projects in Asian cities that serve as bright spots, demonstrating effective implementation models, innovative governance arrangements, and inclusive participatory mechanisms.

Governance & Institutional Bright Spots

Asian countries, and notably China and India, present a diverse range of NbS owing to their vast geographic expanse, socio-ecological diversity, and varying levels of urbanization. Championing the concept of sponge city, China has made substantial investments in green infrastructure and urban greening. Cities such as Beijing, Shenzhen and Shanghai have led large-scale NbS efforts, with notable projects including the Pingshan River Water Treatment Wetlands and the Shenzhen Mangrove Restoration (Qiao et al., 2019).[17]

The policies and frameworks underpinning the Sponge City Initiative in China span interventions from green roofs to urban afforestation programs, and have gained visibility across the region, inspiring similar efforts in neighboring countries. Since its launch with 30 pilot cities in 2015, the program expanded to over 75 cities by 2022 (Yin et al., 2022), positioning it as a central force mainstreaming NbS in Asia. The projects conceived under this policy are reshaping urban landscapes, pioneering the systematic use of NbS in Asia by mitigating the risk of disasters, improving air and water quality, and providing green public spaces.

The Garden City Seoul is a yet another significant NbS initiative led by the Seoul Metropolitan Government aimed at transforming the urban environment through the integration of green spaces, gardens, and ecological infrastructure across the city (Seoul Metropolitan Government, 2023). The initiative is focused on addressing urban environmental challenges such as air pollution, urban heat islands, and flooding by promoting sustainable land use and enhancing the quality

River daylighting in Cheonggyecheon, Seoul (South Korea). Seunghee Han, n.d.

[17] See cases AS_05 and AS_06

of life for residents. At the core of Garden City Seoul is the creation of pocket gardens and the restoration of green corridors that connect parks, rivers, and forests throughout Seoul. These NbS contribute to climate adaptation in dense urban areas, foster biodiversity in the city and increase access to green spaces, mitigating the effects of extreme heat, enhancing air quality, and improving the overall well-being of urban dwellers.

The daylighting of the Cheonggyecheon stream in Seoul was a flagship project which employed a highly participatory and community-focused approach to encourage local ownership and engagement in the design and maintenance of green spaces. Building upon the visibility of this intervention, the city plans to conduct extensive surveys and workshops with residents to identify locations for new gardens and gather input on their design within the Garden City Seoul initiative. Residents are actively involved in the planting of trees and the creation of community gardens, fostering a sense of stewardship and pride in the local environment. One notable feature of the initiative is the integration of ecological design principles, such as the use of rain gardens and permeable surfaces to manage stormwater and reduce urban flooding while creating aesthetically pleasing urban spaces that enhance the livability of neighborhoods.

Terrace Garden at Little Flower School, Adambakkam, Chennai, India. Chennai Resilience Centre, n.d.

ICDS terrace garden in an Anganwadi in Nungambakkam, 2024. Chennai Resilience Centre, n.d.

In India, cities such as Bengaluru, Delhi, Pune, Jaipur, and Chennai are increasingly exploring NbS to address pressing urban challenges including food security, water management, and urban heat islands. One notable example is the Chennai Urban Farming Initiative (Nair and Ayyangar, 2022), which established mobile vegetable gardens in schools, Integrated Child Development Service Centers (ICDS) (anganwadis), and homeless shelters across the city, to enhance climate resilience and foster self-reliant communities.

Kochi, India offers yet another example of transformative urban greening. Under the Kochi Municipal Corporation's Kawaki programme, 12 low-lying sites — previously affected by flooding, waste dumping, and mosquito breeding — have been regenerated as compact urban forests. These areas now host carefully planned native plantings that, once mature, are expected to lower urban heat, improve infiltration through permeable soils, and enhance groundwater recharge.

Cities have invested in flagship projects that have gained regional recognition in the region. For instance, the Bishan Ang Mo Kio Park in Singapore and the Benjakitti Park in Bangkok, stand out for their innovative designs and multifunctional benefits.

As part of the effort, the Corporation engaged workers through the Ayyankali Urban Employment Guarantee Scheme (AUEGS), providing training to over a hundred people. They learned skills such as soil preparation, sapling establishment, irrigation, weeding, and constructing low-cost protective guards – strengthening both ecosystem restoration and local livelihoods (The Better India, 2023).

Elsewhere in Asia, cities have invested in flagship projects that have gained regional recognition in the region. For instance, the Bishan Ang Mo Kio Park in Singapore and the Benjakitti Park in Bangkok, stand out for their innovative designs and multifunctional benefits. Singapore's longstanding commitment to disaster risk reduction and urban greening is exemplified by the ABC Waters Programme, a pioneering effort to integrate NbS into water management (Hamel and Tan, 2022). In 2020, Singapore declared a transition from its earlier 'Garden City' model to a broader strategic vision of becoming 'City in Nature' (Rauf, Hamel, et al., 2024). This shift has led to continued investments in green walls, increased incentives for NbS integration in new developments, and a concerted push to expand urban canopy cover—further reinforcing Singapore's influence on NbS policies throughout the region.

Inclusion and participation Bright Spots

A promising approach in urban NbS relates to co-production approaches in Urban and Peri-urban Agriculture (UPA). The multi-partner Urban and Peri-Urban Agriculture as Green Infrastructure (UPAGrI) project (2019-2022) generated strong evidence on such approaches have strengthened both social well-being and urban sustainability. Conducted across Bengaluru and Pune using qualitative, quantitative, and spatial methods, the research shows that UPA meaningfully enhances urban sustainability through improved indoor thermal comfort, increased wet and dry waste recycling, and widespread graywater reuse. Circular practices – especially waste recycling in cities like Bengaluru and Chennai – demonstrate UPA's critical role in linking composting, food production, and resource recovery. The project also highlights UPA's transformative contributions to societal well-being, including perceived physical and mental health benefits, improved diets, opportunities for income generation, stronger bonds within families and neighborhoods, and the nurturing of pro-environmental values (Koduganti and Patil, 2024).

Another notable bright spot in inclusive and participatory NbS implementation is the Revitalising Informal Settlements and their Environments (RISE) program — an action-research initiative led by a consortium of international universities and implemented across several informal settlements in Makassar, Indonesia.

The RISE program aimed to improve human health and reduce the environmental contamination by using wetlands and biofilters for treating wastewater. Given the challenges of implementing NbS for sanitation in the context of informal settlements, developing a deeply inclusive and participatory co-design approach was a key consideration (Moschonas et al., 2025). Developed over five years, the program conducted several co-design and participatory activities including community mapping exercises and workshops to explain the goals of the NbS. The residents were also invited to participate in a citizen science project, through which scientists and residents collaborated to collect flood data that was later used to design the wetlands.

Construction of wetlands for wastewater treatment in informal settlements within the RISE programme (Indonesia). Erich Wolff, 2019.

Locally-grown plants used in a constructed wetland to treat wastewater within the RISE programme, Makassar (Indonesia). Erich Wolff, 2019.

Between 2018 and 2020, more than 5,000 photos of floods were crowdsourced by residents, offering invaluable information to ensure that the NbS were tailored to the local biophysical conditions. The citizen science project served as a platform to encourage meaningful discussions between residents and researchers, promoting bonds between participants and operating also as a flood early-warning system (Wolff et al., 2021). In later stages of the project, workshops created a safe space for residents to share local concerns with the technology and focus groups with women and children allowed designers to understand the needs of particular groups. Mediated design sessions offered opportunities for the community to identify barriers to implementation and inform the location of specific components of the system. These activities were critical to ensure that the NbS were aligned with the local aspirations and priorities (Moschonas et al., 2025). Results have shown that the co-design approach played an important role by bridging the divide between NbS experts and communities, disseminating knowledge about the NbS, as well as increasing social capital and strengthening community bonds (Salinger et al., 2024).

In other contexts, innovative participatory projects in Asia offered important insights into how NbS can be implemented using a context-sensitive approach that positions communities at the center of climate adaptation debates. A genuine commitment to inclusivity in NbS requires recognizing and valuing diverse ways of knowing, including Indigenous and local knowledge systems, and respecting different value frameworks (Chan et al., 2016; Mabon, 2019; Wolff et al., 2025). Examples of more inclusive approaches include ProCoast's Coastal Adaptation Initiatives in the Philippines,[18] and Thailand's 'Nature in the City' approach. Both of these endeavors employed participatory planning workshops that brought together residents, businesses, and government officials to collaboratively design and implement NbS solutions. In the case of ProCoast, these partnerships play a key role by establishing participatory monitoring and management strategies that can adaptively respond to the ongoing challenges of mainstreaming mangrove restoration.

At a policy level, a growing number of Asian countries are developing national strategies for climate adaptation and implementation of urban NbS. Regional collaborations, such as programs to invest in biodiversity recently announced between ASEAN and the European Union highlight a growing interest in advancing NbS in the region. Building upon local experiences with water sensitive urban drainage, natural solutions and climate resilience initiatives, regional collaborations will be critical to mainstream NbS in the region by disseminating knowledge on locally developed systems across national borders. Robust policies are going to be critical to ensure that NbS are employed to protect key ecosystems and contribute to environmental justice in new cities, such as Nusantara in Indonesia (Yuanita et al., 2025) and Putrajaya in Malaysia.[19]

[18] *See cases AS_11*

[19] *See cases AS_15*

A genuine commitment to inclusivity in NbS requires recognizing and valuing diverse ways of knowing, including Indigenous and local knowledge systems, and respecting different value frameworks.

The mangrove restoration project ProCoast is protecting waterfront areas in Iloilo using a participatory approach (the Philippines). Arnel Murga, 2025.

Maha Bandula Park in Yangon (Myanmar) provides shade in the centre of the city. Chandni Singh, 2017.

4 Recommendations and Future Directions

Urban greening improves liveability but has also been used to increase property value in commercial developments. Bonifacio High Street, Taguig, Metro Manila (The Philippines). Erich Wolff, 2025.

The Asia Roadmap for Urban Nature-based Solutions highlights the widespread implementation of NbS across Asian cities, spanning diverse ecosystems, actors, governance strategies, and challenges. However, to effectively adopt and implement NbS in supporting climate action, biodiversity conservation, and urban resilience, strategies must be tailored to the unique urbanization patterns and socio-political contexts shaping cities across the region. The following recommendations outline key priorities for advancing NbS in Asian cities:

4.1 Recommendations

KEY RECOMMENDATION 1:

Strengthening monitoring, evaluation and learning of different types of NbS in Asian contexts will accelerate uptake and build a robust evidence base and benchmarks.

There is an urgent need for context-sensitive, robust and systematic monitoring and evaluation efforts to assess the effectiveness of NbS in achieving biodiversity and climate action goals. Despite promising outcomes—such as reduced flooding and mitigation of urban heat—empirical evidence on NbS performance across spatial and temporal scales remains limited (Tun et al., 2024). Many flagship initiatives are still limited to demonstration sites, and monitoring and evaluation frameworks remain underdeveloped. This results in gaps in tracking progress, identifying barriers, and demonstrating value to policymakers and investors.

Building regional databases, documenting lessons from flagship projects, and investing in technical expertise will help create benchmarks to guide future implementation. Monitoring approaches must also capture both co-benefits and trade-offs, acknowledging that NbS can generate disservices alongside positive outcomes (Lechner et al., 2021; Shaw et al., 2022). Integrating low-cost monitoring technologies and community-driven methods — such as citizen science for tracking biodiversity, water levels, or water quality — offers scalable pathways to strengthen data availability, enhance ownership, and support evidence-informed decision-making across the region.

KEY RECOMMENDATION 2:

Integrating NbS into urban planning is essential for advancing climate and biodiversity goals.

Mainstreaming NbS into urban planning is essential for aligning climate and biodiversity objectives with spatial development strategies. While some progress has been made in this direction, existing urban policies across Asia often fail to acknowledge the intersections between NbS and related domains such as housing, transport, infrastructure, recreation, and cultural heritage (Rauf, Wolff, et al., 2024). Targeted NbS strategies must address challenges specific to the region, including the urban-peri-urban-rural interface, rapid growth of informal settlements, and population movements due to regional, national and international migration. Greater policy integration will enable cities to position NbS as central tools for climate adaptation and inclusive urban development.

Waterways are a key part of daily life. The River Aerawaddy in Myanmar is used for transport, leisure, and central to local religious life. Chandni Singh, 2018.

KEY RECOMMENDATION 3:

Developing diversified funding mechanisms, flexible governance models, and robust evaluation systems is essential for effectively mainstreaming NbS across scales in Asia — ensuring their effectiveness under diverse climatic, ecological, and governance conditions.

Successfully mainstreaming NbS across Asia requires diversified financing and governance mechanisms that can simultaneously support large-scale infrastructure projects and smaller, community-led initiatives. While national governments and international funders currently provide substantial backing for major projects (Lechner et al., 2020), small-scale and grassroots NbS often struggle to access funding, technical assistance, and institutional recognition. Expanding innovative financing options such as micro-grants, climate adaptation funds, and community co-investment models can help scale up these local initiatives and strengthen inclusion. Blended finance mechanisms, where public funds help de-risk private investment, also offer a promising pathway to increase corporate engagement in NbS.

At the same time, there is a growing need to articulate 'financeable NbS options by city typology.' Cities differ widely in climate risks, population size, and development patterns, and therefore require tailored NbS solutions. A clearly defined menu of context-appropriate interventions, with indicative costs and expected benefits, would allow funders to make more informed investment decisions and accelerate NbS uptake across diverse urban settings.

To complement these financing and governance reforms, more research is needed to evaluate the effectiveness of NbS across varied ecological, climatic, and governance contexts. This includes developing practical, context-sensitive tools and metrics that assess not only ecological impact but also social relevance, cost-effectiveness, and long-term feasibility. NbS must be supported as holistic interventions, backed by resources for long-term maintenance and capacity-building programs, that equip communities to sustain projects beyond initial implementation. Without such supportive mechanisms and robust evaluation frameworks, even well-designed NbS risk underperformance or abandonment.

KEY RECOMMENDATION 4:

Fostering inclusive participation and integrating Indigenous and local knowledge is crucial to ensure context-relevant and socially legitimate NbS.

Meaningful collaboration with communities — especially those living in informal settlements— is foundational for the success, ownership, and long-term sustainability of urban NbS. Actively involving residents in the design, planning, implementation, and monitoring of projects fosters trust, encourages experimentation, and creates opportunities for co-learning. At the same time, embedding Indigenous, local, and traditional ecological knowledge is critical for ensuring that NbS remain context-relevant, equitable, and responsive to historic socio-environmental inequalities across Asian cities (Salim, Bettinger and Fisher, 2019). Empowering communities through participatory mechanisms, capacity-building programs, and support systems that strengthen their engagement with governmental and institutional actors can enhance social legitimacy, integrate culturally rooted values, and ultimately improve the effectiveness and uptake of NbS interventions.

Model used for community engagement in the RISE program (Indonesia), developed by Dasha Moschonas. Erich Wolff, 2019.

Participatory mapping exercise to plan nature-based solutions in Rangsit (Thailand). Erich Wolff, 2023.

KEY RECOMMENDATION 5:

Strengthening context-sensitive NbS by addressing regional gaps and advancing locally grounded frameworks for smaller cities.

Research and practice on NbS in Asia remain uneven, with higher-income countries such as Singapore and China generating more peer-reviewed evidence, policy innovations, and flagship projects. This disparity limits the ability of lower-income countries, especially secondary, smaller, and rapidly urbanizing peri-urban centers to develop context-specific solutions or adapt global best practices effectively. To widen adoption across the region, it is essential to build capacity, generate locally relevant knowledge, and document how NbS perform across diverse socio-ecological settings ranging from peri-urban informal settlements, to urban rivers, to large urban parks.

At the same time, developing locally grounded, context-sensitive frameworks is critical for understanding and evaluating NbS beyond megacities. Smaller cities and peri-urban areas offer unique opportunities for experimentation, enabling practitioners to identify regional trends and design frameworks that reflect local priorities, governance realities, and community practices. Such frameworks must balance local ingenuity with external lessons, avoiding both over-reliance on Global North models and romanticization of vernacular approaches. Strengthening regional knowledge-sharing platforms and collaborative databases will help generate comparable case studies, support peer learning, and build a more equitable, robust evidence base for effective NbS strategies across Asia.

KEY RECOMMENDATION 6:

Advancing NbS implementation across Asia requires strong regional collaboration, pooled technical and research capacities, and structured knowledge & innovation exchange via platforms like Global Covenant of Mayors to accelerate evidence-based and climate resilient interventions.

Fostering regional collaboration and cross-border innovation exchange is essential for advancing context-relevant NbS strategies across Asia. Higher-income countries can play a pivotal role by investing in shared research, technical assistance, and capacity-building initiatives that strengthen the regional evidence base and enable mutual learning across diverse socio-economic and ecological contexts. Collaborative research platforms can serve as engines of sustainable development, generating locally grounded yet transferable models of NbS that support climate resilience and biodiversity conservation across the region. Experiences from initiatives such as the Global Covenant of Mayors (GCoM) and other transnational city networks further illustrate how structured knowledge-sharing mechanisms can accelerate adoption and scale-up of effective NbS practices.

Asih Busiadi, GCoM Asia Team Leader and Thunpicha Pinniam, GCoM Asia Project Assistant. DAI.com for the European Union, 2022.

4.2 Limitations and Future Research Pathways

This report provides an overview of the state of urban NbS research and practice across Asia. It synthesizes available evidence from across the region but given the diversity and breadth of Asia it does not detail the multiple realities, priorities and nuances across the continent. Accounting for the limitations of this report and proposing ways to bridge the gaps in the field, we distil future research directions.

First, urban NbS research and documentation across Asia remain uneven. Higher-income countries, such as Singapore and China, are at the forefront in peer-reviewed publications, policy development, and flagship implementation. In contrast, many lower-income countries struggle to generate context-specific evidence or adapt global best practices effectively. As a result, the regional evidence base is skewed toward countries with greater resources, potentially obscuring innovative but less documented NbS practices in secondary cities, smaller urban centers, and locally driven initiatives.

Compounding this imbalance is the limited availability of data on community-led or small-scale projects. The focus on large, well-funded NbS projects—primarily in East, South and Southeast Asia—means that many locally-led practices, especially those with inadequate documentation or long-term monitoring, evaluation and learning, are overlooked. Future research should prioritize dedicated investment in strengthening monitoring, evaluation and learning frameworks and improving the communication of ecological, social, and economic outcomes of urban NbS across diverse local contexts.

Second, the team encountered difficulties in interpreting some documents describing urban NbS in the region. Many regional reports, case studies, and practitioner documents are

Urban NbS research and documentation across Asia remain uneven. As a result, the regional evidence base is skewed toward countries with greater resources, potentially obscuring innovative but less documented NbS practices in secondary cities, smaller urban centers, and locally driven initiatives.

Developing regional knowledge-sharing platforms and open-access databases will enable local governments, researchers, and civil society to collaborate, experiment, and generate comparable case studies.

published in local languages or behind restricted access, leading to partial underrepresentation of important examples. The team used all resources available, ranging from translation tools to consultations with local experts to synthesize knowledge in the region but some restrictions to the access to country-specific knowledge still persist. The team consisted of South and Southeast Asian experts, which may have influenced interpretive balance across subregions. However, care was taken to identify and include cases from Central, East, and West Asia. Future work should emphasize multilingual approaches and more regionally diverse collaborations to enhance the inclusiveness of synthesized evidence.

Finally, limited availability of regional databases and regional cooperation mechanisms for urban NbS in Asia created challenges for developing a strategy to review, conceptualize and synthesize comparable evidence across the region. This barrier is also reflected in the field more broadly, as stronger multi-partner and multi-regional collaboration are still required for advancing NbS in Asia. To support wider adoption — particularly in secondary and smaller cities — capacity-building efforts must focus on documenting how NbS can be adapted to the socio-ecological and spatial diversity of urban contexts. Developing regional knowledge-sharing platforms and open-access databases will enable local governments, researchers, and civil society to collaborate, experiment, and generate comparable case studies. Higher-income countries should continue investing in cross-border research and technical assistance to strengthen the regional evidence base. Such collaborative research platforms can drive innovation, mutual learning, and the development of locally grounded yet transferable NbS models for climate resilience and biodiversity conservation.

5 References

1. Abbu, N. et al. (2015) Urban Green Growth Strategies for Indian Cities. Vol. 1. Delhi, India: ICLEI - Local Governments for Sustainability, South Asia. Available at: https://southasia.iclei.org/wp-content/uploads/2021/09/UrbanGreenGrowth.pdf.

2. ACCCRN (2014) ACCCRN City Projects. Bangkok, Thailand: The Rockefeller Foundation, Asia Office. Available at: https://acccrn.net/sites/default/files/publication/attach/ACCCRN%20Cities%20Project_June%202014.pdf.

3. Adambekov, S. et al. (2016) "Health challenges in Kazakhstan and Central Asia," Journal of Epidemiology and Community Health, 70(1), pp. 104–108. Available at: https://doi.org/10.1136/jech-2015-206251.

4. Andrews, B. et al. (2023) The Critical Role of Nature-based Solutions for Enhancing Climate Resilience in Informal Areas: An Urban Supplement to the UNFCCC Technical Guidelines on National Adaptation Plans. Nairobi: United Nations Human Settlements Programme (UN-Habitat). Available at: https://unhabitat.org/the-critical-role-of-nature-based-solutions-for-enhancing-climate-resilience-in-informal-areas.

5. Anguelovski, I. et al. (2016) "Equity Impacts of Urban Land Use Planning for Climate Adaptation: Critical Perspectives from the Global North and South," Journal of Planning Education and Research, 36(3), pp. 333–348. Available at: https://doi.org/10.1177/0739456X16645166.

6. Archer, D. et al. (2020) "The role of collective and individual assets in building urban community resilience," International Journal of Urban Sustainable Development, 12(2), pp. 169–186. Available at: https://doi.org/10.1080/19463138.2019.1671425.

7. Ashraf, U. (2020) Green, Green Rocky Road: The Perils of Greenwashing Pakistan, Jamhoor. Available at: https://www.jamhoor.org/read/2020/9/6/green-green-rocky-road-the-perils-of-greenwashing-in-pakistan.

8. Asian Development Bank (2019) Nature-Based Solutions for Cities in Viet Nam: Water Sensitive Urban Design. Asian Development Bank. Available at: https://www.adb.org/publications/nature-based-solutions-cities-viet-nam.

9. Asian Development Bank (2020) Asian Water Development Outlook 2020: Advancing Water Security across Asia and the Pacific. Asian Development Bank. Available at: https://www.adb.org/publications/asian-water-development-outlook-2020 (Accessed: November 3, 2025).

10. Cao, A. et al. (2021) "Future of Asian Deltaic Megacities under sea level rise and land subsidence: current adaptation pathways for Tokyo, Jakarta, Manila, and Ho Chi Minh City," Current Opinion in Environmental Sustainability, 50, pp. 87–97. Available at: https://doi.org/10.1016/j.cosust.2021.02.010.

11. Castaldo, A.G. et al. (2025) "Nature-based solutions and urban planning in the Global South: Challenge orientations, typologies, and viability for cities," Land Use Policy, 150, p. 107439. Available at: https://doi.org/10.1016/j.landusepol.2024.107439.

12. Chan, K.M.A. et al. (2016) "Why protect nature? Rethinking values and the environment," Proceedings of the National Academy of Sciences, 113(6), pp. 1462–1465. Available at: https://doi.org/10.1073/pnas.1525002113.

13. Chatterjee, M. (2010) "Slum dwellers response to flooding events in the megacities of India," Mitigation and Adaptation Strategies for Global Change, 15(4), pp. 337–353. Available at: https://doi.org/10.1007/s11027-010-9221-6.

14. Cohen-Shacham, E. et al. (eds.) (2016) Nature-based solutions to address global societal challenges. IUCN International Union for Conservation of Nature. Available at: https://doi.org/10.2305/IUCN.CH.2016.13.en.

15. Dabla-Norris, E., Nozaki, M. and Daniel, J. (2021) Asia's Climate Emergency, IMF. Available at: https://www.imf.org/en/Publications/fandd/issues/2021/09/asia-climate-emergency-role-of-fiscal-policy-IMF-dabla (Accessed: November 3, 2025).

16. Dangol, N. and Day, J. (2017) "Flood adaptation by informal settlers in kathmandu and their fear of eviction," International Journal of Safety and Security Engineering, 7(2), pp. 147–156. Available at: https://doi.org/10.2495/SAFE-V7-N2-147-156.

17. Dodman, D. et al. (2023) "Cities, Settlements and Key Infrastructure," in Climate Change 2022 – Impacts, Adaptation and Vulnerability: Working Group II Contribution to the Sixth Assessment Report of the Intergovernmental

Panel on Climate Change. 1st ed. Cambridge, UK and New York, NY, USA: Cambridge University Press, pp. 907–1040. Available at: https://doi.org/10.1017/9781009325844.

18. Dumitru, A., Frantzeskaki, N. and Collier, M. (2020) "Identifying principles for the design of robust impact evaluation frameworks for nature-based solutions in cities," Environmental Science & Policy, 112, pp. 107–116. Available at: https://doi.org/10.1016/j.envsci.2020.05.024.

19. Erbacher, A. and Wilkens, M. (2024) Increasing climate resilience in ASEAN cities through inclusive Nature-based Solutions. Developing projects to empower women and other people in vulnerable situations. Berlin: adelphi research gGmbH. Available at: https://adelphi.de/en/publications/increasing-climate-resilience-in-asean-cities-through-inclusive-nature-based-solutions.

20. Firdaus, F.M., Elliott, B. and Ibanez, D. (2023) Southeast Asian Cities Have Some of the Most Polluted Air in the World. El Niño Is Making it Worse, World Resources Institute. Available at: https://www.wri.org/insights/air-pollution-southeast-asia-cities-jakarta-el-nino.

21. Gadgil, G.U. and Baker, J.L. (2017) East Asia and Pacific Cities: Expanding Opportunities for the Urban Poor. World Bank, Washington, DC. Available at: https://doi.org/10.1596/978-1-4648-1093-0.

22. Ghazi, S. (2024) "Rise of new megacities will drive global urban growth," Oxford Economics, 26 January. Available at: https://www.oxfordeconomics.com/resource/rise-of-new-megacities-will-drive-global-urban-growth/.

23. Greenwalt, J. et al. (2023) Strategy Paper on Nature-based Solutions to Build Climate Resilience in Informal Areas. Nairobi: United Nations Human Settlements Programme (UN-Habitat). Available at: https://unhabitat.org/strategy-paper-on-nature-based-solutions-to-build-climate-resilience-in-informal-areas.

24. Hamel, P. and Tan, L. (2022) "Blue–Green Infrastructure for Flood and Water Quality Management in Southeast Asia: Evidence and Knowledge Gaps," Environmental Management, 69(4), pp. 699–718. Available at: https://doi.org/10.1007/s00267-021-01467-w.

25. Hong, T. et al. (2021) "Government-driven urbanisation and its impact on regional economic growth in China," Cities, 117, p. 103299. Available at: https://doi.org/10.1016/j.cities.2021.103299.

26. Irvine, K.N. et al. (2023) "Nature-based solutions to enhance urban flood resiliency: case study of a Thailand Smart District," Sustainable Water Resources Management, 9(1), p. 43. Available at: https://doi.org/10.1007/s40899-023-00821-6.

27. Irvine, K.N., Ho, H.L. and Chua, L.H.C. (2021) "Dynamics of runoff quality associated with an urban park and WSUD treatment train in a tropical climate," Environmental Technology, 44(4), pp. 512–527. Available at: https://doi.org/10.1080/09593330.2021.1976844.

28. Iwachido, Y. et al. (2020) "Nature-oriented park use of satoyama ecosystems can enhance biodiversity conservation in urbanized landscapes," Landscape and Ecological Engineering, 16(2), pp. 163–172. Available at: https://doi.org/10.1007/s11355-020-00413-y.

29. Khan, M.R. and Munira, S. (2021) "Climate change adaptation as a global public good: implications for financing," Climatic Change, 167(3–4), p. 50. Available at: https://doi.org/10.1007/s10584-021-03195-w.

30. Kharisma, D. and Kaswanto (2021) "Management of home garden in formal and informal settlement along riparian Ciliwung river in Bogor," IOP Conference Series: Earth and Environmental Science, 694(1), p. 012021. Available at: https://doi.org/10.1088/1755-1315/694/1/012021.

31. Koduganti, M. and Patil, S. (2024) "Cultivating Change in Cities: Exploring and Classifying the Determinants of Urban Agriculture in India," in Sustainable Urban Agriculture. CRC Press.

32. KPMG China (2016) The 13th Five-Year Plan – China's transformation and integration with the world economy: Opportunities for Chinese and foreign businesses. Available at: https://assets.kpmg.com/content/dam/kpmg/cn/pdf/en/2016/10/13fyp-opportunities-analysis-for-chinese-and-foreign-businesses.pdf.

33. Lakshmisha, A., Nazar, A.F. and Nagendra, H. (2024) "Nature based solutions in cities of the global South—The 'where, who and how' of implementation," Environmental Research: Ecology, 3(2), p. 025005. Available at: https://doi.org/10.1088/2752-664X/ad53cf.

34. Lechner, A.M. et al. (2020) "Challenges and considerations of applying nature-based solutions in low- and middle-income countries in Southeast and East Asia," Blue-Green Systems, 2(1), pp. 331–351. Available at: https://doi.org/10.2166/bgs.2020.014.

35. Lechner, A.M. et al. (2021) Urban Biodiversity and Nature-Based Solutions in Southeast Asia: Perspectives from Indonesia and Malaysia. ISEAS–Yusof Ishak Institute Singapore. Available at: https://doi.org/10.1355/9789815011210.

36. Likitswat, F. and Sung, C. (2025) "Implementing Nature-based Solutions: Learning from Benchakitti Forest Park in Bangkok," Journal of Architectural/Planning Research and Studies (JARS), 22(2), p. 272299. Available at: https://doi.org/10.56261/jars.v22.272299.

37. Linh, N.S., Ahmed, F. and Loc, H.H. (2022) "Applications of Nature-Based Solutions in Urban Water Management in Singapore, Thailand and Vietnam: A Review," in N. Pachova et al. (eds.) Regional Perspectives of Nature-based Solutions for Water: Benefits and Challenges. Cham: Springer International Publishing (Applied Environmental Science and Engineering for a Sustainable Future), pp. 101–126. Available at: https://doi.org/10.1007/978-3-031-18412-3_5.

38. Luo, M. et al. (2023) Towards Nature-based Solutions at scale: 10 case studies from China. Gland, Switzerland: IUCN, and Beijing, the People's Republic of China: Ministry of Natural Resources. Available at: https://portals.iucn.org/library/sites/library/files/documents/2023-012-En.pdf.

39. Mabon, L. (2019) "Enhancing post-disaster resilience by 'building back greener': Evaluating the contribution of nature-based solutions to recovery planning in Futaba County, Fukushima Prefecture, Japan," Landscape and Urban Planning, 187, pp. 105–118. Available at: https://doi.org/10.1016/j.landurbplan.2019.03.013.

40. Matthews, J. and Dela Cruz, E.O. (2022) Integrating Nature-Based Solutions for Climate Change Adaptation and Disaster Risk Management: A Practitioner's Guide. Asian Development Bank. Available at: https://doi.org/10.22617/TIM220215-2.

41. McGee, T.G. (1991) "Chapter 1 The Emergence of Desakota Regions in Asia: Expanding a Hypothesis," in N. Ginsburg, B. Koppel, and T.G. McGee (eds.) The Extended Metropolis. University of Hawaii Press, pp. 1–26. Available at: https://doi.org/10.1515/9780824841904-005.

42. Mollan, C. (2024) Delhi air pollution levels at "severe plus", authorities say, BBC News. Available at: https://www.bbc.com/news/articles/c1wj43vqdlpo.

43. Møller, K.A. et al. (2012) "Economic, environmental and socio-cultural sustainability of three constructed wetlands in Thailand," Environment and Urbanization, 24(1), pp. 305–323. Available at: https://doi.org/10.1177/0956247811434259.

44. Moschonas, D. et al. (2025) "Co-producing nature-based solutions: integrating knowledge interfaces in informal settlements in Indonesia," Environment & Urbanization, 37(2), pp. 465–486. Available at: https://doi.org/10.1177/09562478251364223.

45. Mukhopadhyay, S. and Chowdhury, S. (2021) [Commentary] Nature-based solutions hold the key to transform our cities into resilient spaces, Mongabay. Available at: https://india.mongabay.com/2021/07/commentary-nature-based-solutions-hold-the-key-to-transform-our-cities-into-resilient-spaces/.

46. Musadat, A. (2019) Participatory Planning and Budgeting in Decentralised Indonesia: Understanding Participation, Responsiveness and Accountability. Australian National University. Available at: https://openresearch-repository.anu.edu.au/bitstreams/25d6a91f-059e-49ff-8685-da506e68b55c/download.

47. Nair, R.M. and Ayyangar, A. (2022) "Building Resilient Food Systems: The Chennai Urban Farming Initiative," Resilient Cities Network. Available at: https://resilientcitiesnetwork.org/building-resilient-food-systems-the-chennai-urban-farming-initiative/.

48. Nassary, E.K. et al. (2022) "Exploring urban green packages as part of Nature-based Solutions for climate change adaptation measures in rapidly growing cities of the Global South," Journal of Environmental Management, 310, p. 114786. Available at: https://doi.org/10.1016/j.jenvman.2022.114786.

49. Nesshöver, C. et al. (2017) "The science, policy and practice of nature-based solutions: An interdisciplinary perspective," Science of The Total Environment, 579, pp. 1215–1227. Available at: https://doi.org/10.1016/j.scitotenv.2016.11.106.

50. Office of the Registrar General & Census Commissioner, India (2011) Census of India 2011. New Delhi: Ministry of Home Affairs, Government of India.

51. Patil, S. et al. (Under review) "Urban agriculture in India: mapping its landscape and sustainability and well-being outcomes," Local Environment [Preprint].

52. Pauleit, S. et al. (2017) "Nature-Based Solutions and Climate Change – Four Shades of Green," in N. Kabisch et al. (eds.) Nature-Based Solutions to Climate Change Adaptation in Urban Areas. Cham: Springer International Publishing (Theory and Practice of Urban Sustainability Transitions), pp. 29–49. Available at: https://doi.org/10.1007/978-3-319-56091-5_3.

53. Pottinger-Glass, C. (2022) Nature-based solutions are no quick fix for making cities more liveable, Stockholm Environment Institute. Available at: https://www.sei.org/perspectives/nature-based-solutions-are-no-quick-fix-for-making-cities-more-liveable/.

54. Prana, A.M. and Susilo, F.S.B.P. (2020) Jakarta flood management requires paradigm shift from "control" to "adaptation" - Opinion, The Jakarta Post. Available at: https://www.thejakartapost.com/academia/2020/03/16/jakarta-flood-management-requires-paradigm-shift-from-control-to-adaptation.html.

55. Qiao, X.-J. et al. (2019) "Governance factors of sustainable stormwater management: A study of case cities in China and Sweden," Journal of Environmental Management, 248, p. 109249. Available at: https://doi.org/10.1016/j.jenvman.2019.07.020.

56. Rao, N. et al. (2022) "Cultivating sustainable and healthy cities: A systematic literature review of the outcomes of urban and peri-urban agriculture,"

Sustainable Cities and Society, 85, p. 104063. Available at: https://doi.org/10.1016/j.scs.2022.104063.

57. Rauf, H.A., Wolff, E., et al. (2024) "Aligning nature-based solutions and housing policy: A study on stakeholders' perceptions of nature within informal settlement upgrading projects," Cities, 153, p. 105264. Available at: https://doi.org/10.1016/j.cities.2024.105264.

58. Rauf, H.A., Hamel, P., et al. (2024) "Incorporating tiny forests in the neighbourhood of Singapore - Opportunities and challenges," Nature-Based Solutions, 6, p. 100193. Available at: https://doi.org/10.1016/j.nbsj.2024.100193.

59. Salim, W., Bettinger, K. and Fisher, M. (2019) "Maladaptation on the Waterfront: Jakarta's Growth Coalition and the Great Garuda," Environment and Urbanization ASIA, 10(1), pp. 63–80. Available at: https://doi.org/10.1177/0975425318821809.

60. Salinger, A.P. et al. (2024) "'People are now working together for a common good': The effect on social capital of participatory design for community-level sanitation infrastructure in urban informal settlements," World Development, 174, p. 106449. Available at: https://doi.org/10.1016/j.worlddev.2023.106449.

61. Sato, M., Minamoto, T. and Ushimaru, A. (2023) "An interdisciplinary approach to environmental conservation policy: a case of Satoyama redevelopment in the peri-urban area," International Journal of Economic Policy Studies, 17(2), pp. 403–419. Available at: https://doi.org/10.1007/s42495-023-00110-2.

62. Seddon, N. et al. (2020) "Understanding the value and limits of nature-based solutions to climate change and other global challenges," Philosophical Transactions of the Royal Society B: Biological Sciences, 375(1794), p. 20190120. Available at: https://doi.org/10.1098/rstb.2019.0120.

63. Seddon, N. et al. (2021) "Getting the message right on nature-based solutions to climate change," Global Change Biology, 27(8), pp. 1518–1546. Available at: https://doi.org/10.1111/gcb.15513.

64. Seoul Metropolitan Government (2023) Seoul's roadmap to expand the city's urban green spaces. Available at: https://english.seoul.go.kr/seouls-roadmap-to-expand-the-citys-urban-green-spaces/.

65. Shaikh, S.F.E.A. and Hamel, P. (2023) "Identifying nature-positive futures in new cities: an application of the Urban Nature Futures Framework," Sustainability Science [Preprint]. Available at: https://doi.org/10.1007/s11625-023-01411-3.

66. Shannon, K. and Legrand, B. (2007) Aerated Lagoon Park in Ho Chi Minh City, Vietnam. 59. Topos: European Landscape Magazine, pp. 31–37. Available at: https://lirias.kuleuven.be/retrieve/498381.

67. Shaw, R. et al. (2022) Climate Change 2022 – Impacts, Adaptation and Vulnerability: Working Group II Contribution to the Sixth Assessment Report of the Intergovernmental Panel on Climate Change. 1st ed. Cambridge University Press. Available at: https://doi.org/10.1017/9781009325844.

68. Singh, C., Madhavan, M., et al. (2021) "Climate change adaptation in Indian cities: A review of existing actions and spaces for triple wins," Urban Climate, 36, p. 100783. Available at: https://doi.org/10.1016/j.uclim.2021.100783.

69. Singh, C., Jain, G., et al. (2021) "Losses and damages associated with slow-onset events: urban drought and water insecurity in Asia," Current Opinion in Environmental Sustainability, 50, pp. 72–86. Available at: https://doi.org/10.1016/j.cosust.2021.02.006.

70. Singh, C., Koduganti, M. and Patil, S. (2023) "Leading from the Top: Wet Waste Composting and Rooftop Gardening in Pune Municipality Offices," in Sowing Sustainable Cities: Lessons for Urban Agriculture Practices in India. Indian Institute for Human Settlements. Available at: https://doi.org/10.24943/SSC12.2023.

71. Singh, C., Koduganti, M. and Patil, S. (2025) "Does urban agriculture make us healthier and happier? Examining well-being outcomes of home gardening in India."

72. Sui, D.Z. and Zeng, H. (2001) "Modeling the dynamics of landscape structure in Asia's emerging desakota regions: a case study in Shenzhen," Landscape and Urban Planning, 53(1), pp. 37–52. Available at: https://doi.org/10.1016/S0169-2046(00)00136-5.

73. Tan, S.B. and Low, D. (2019) "Re-Examining Singapore's Urban Planning and Governance Framework," in S. Hamnett and B. Yuen (eds.) Planning Singapore: The Experimental City. 1st ed. Routledge. Available at: https://doi.org/10.4324/9781351058230.

74. Thakur Nanda, S. (2025) Public private partnerships build urban forests, Mongabay. Available at: https://india.mongabay.com/2025/02/ppp-forests-partnerships-for-planet/.

75. The Better India (2023) 3 Nature-Based Solutions That Are Making Low-Income Communities Resilient to Climate Change. Available at: https://thebetterindia.com/328522/how-nature-based-solutions-can-help-build-climate-change-resilience-wri-projects/.

76. The Green Bein (2025) Solar Energy Property Tax Benefits in Maharashtra & Goa, The Green Bein. Available at: https://www.thegreenbein.in/post/solar-energy-property-tax-benefits-in-maharashtra.

77. Thompson, A. et al. (2023) "Paying for nature-based solutions: A review of funding and financing mechanisms for ecosystem services and their impacts on social equity," Sustainable Development, 31(4), pp.

1991–2066. Available at: https://doi.org/10.1002/sd.2510.

78. Tun, K.Z. et al. (2024) "Mainstreaming Nature-Based Solutions for Climate Adaptation in Southeast Asia: A Systematic Review," Earth Systems and Environment, 9(2), pp. 1333–1351. Available at: https://doi.org/10.1007/s41748-024-00473-2.

79. UN ESCAP (2025) Urban transformation in Asia and the Pacific: from growth to resilience. Edited by Vereinte Nationen. Bangkok: Economic and Social Commission for Asia and the Pacific.

80. United Nations (2017) "World Urbanization Prospects." Available at: https://population.un.org/wup/.

81. United Nations Development Programme (2023) Bhutan secures $8.9 million from GEF Least Developed Countries Fund towards a more climate-resilient water sector. Available at: https://www.adaptation-undp.org/bhutan-secures-89-million-gef-least-developed-countries-fund-towards-more-climate-resilient-water.

82. United Nations Environment Programme (2023) Nature-based Infrastructure: How natural infrastructure solutions can address sustainable development challenges and the triple planetary crisis. United Nations Environment Programme. Available at: https://doi.org/10.59117/20.500.11822/44022.

83. United Nations Human Settlements Programme (2020) World Cities Report 2020: The Value of Sustainable Urbanization. Erscheinungsort nicht ermittelbar: United Nations.

84. Urban Nature Atlas (2024) Integrated Development of the Hatirjheel Area. Available at: https://una.city/nbs/dhaka/integrated-development-hatirjheel-area.

85. Urban Poor Development Fund and Asian Coalition for Housing Rights (2010) Citywide Upgrading in 11 Cambodian Cities. Cambodia. Available at: https://urbandatabase.khmerstudies.org/get-data/MTQ1/view.

86. Van Raalte, D. and Ranger, N. (2023) Financing Nature-Based Solutions for Adaptation at Scale: Learning from Specialised Investment Managers and Nature Funds. Global Center on Adaptation and Environmental Change Institute, University of Oxford. Available at: https://www.eci.ox.ac.uk/sites/default/files/2023-12/Financing_NbS_for_Adaptation-GCAOxford2023-finalv2.pdf.

87. Wolff, E. et al. (2021) "Collaborating With Communities: Citizen Science Flood Monitoring in Urban Informal Settlements," Urban Planning, 6(4), pp. 351–364. Available at: https://doi.org/10.17645/up.v6i4.4648.

88. Wolff, E. et al. (2025) "Nature for the people, by the people: Negotiating values, attitudes and behaviours for implementing urban nature-based solutions in social housing," Urban Forestry & Urban Greening, 112, p. 128973. Available at: https://doi.org/10.1016/j.ufug.2025.128973.

89. Wolff, E., Rauf, H.A. and Hamel, P. (2023) "Nature-based solutions in informal settlements: A systematic review of projects in Southeast Asian and Pacific countries," Environmental Science & Policy, 145, pp. 275–285. Available at: https://doi.org/10.1016/j.envsci.2023.04.014.

90. World Bank (2023a) Central Asia Resilient and Low-carbon Cities (CARL-cities). Available at: https://thedocs.worldbank.org/en/doc/09a6e5c1e-53c4d9be5739c02266cb4ba- 0080012023/original/Fact-Sheet-CARL-Cities-2023-en.pdf.

91. World Bank (2023b) Cities Across Central Asia Can Unlock Full Economic Potential by Implementing Low-Carbon and Climate-Resilient Development Strategies, World Bank. Available at: https://www.worldbank.org/en/news/press-release/2023/09/27/cities-across-central-asia-can-unlock-full-economic-potential-by-implementing-low-carbon-development-strategies.

92. Yarina, L. (2018) "Your Sea Wall Won't Save You: Negotiating Rhetorics and Imaginaries of Climate Resilience," Places Journal [Preprint], (2018). Available at: https://doi.org/10.22269/180327.

93. Yin, D. et al. (2022) "Sponge City Practices in China: From Pilot Exploration to Systemic Demonstration," Water, 14(10), p. 1531. Available at: https://doi.org/10.3390/w14101531.

94. Yuanita, C.N. et al. (2025) "Nature-based Solutions for Hydrometeorological Risk Reduction: The Challenges and Opportunities of Riparian Forest in Indonesia's New Capital City," IOP Conference Series: Earth and Environmental Science, 1447(1), p. 012017. Available at: https://doi.org/10.1088/1755-1315/1447/1/012017.

95. Zhang, J. et al. (2021) "Climate Change Decreased Net Ecosystem Productivity in the Arid Region of Central Asia," Remote Sensing, 13(21), p. 4449. Available at: https://doi.org/10.3390/rs13214449.

96. Zhang, Z. et al. (2024) "Municipal solid waste management challenges in developing regions: A comprehensive review and future perspectives for Asia and Africa," Science of The Total Environment, 930, p. 172794. Available at: https://doi.org/10.1016/j.scitotenv.2024.172794.

97. Zhu, A.L. et al. (2024) "China's nature-based solutions in the Global South: Evidence from Asia, Africa, and Latin America," Global Environmental Change, 86, p. 102842. Available at: https://doi.org/10.1016/j.gloenvcha.2024.102842.

Visual References

1. Singapore's Bishan-Ang Mo Kio Park. Erich Wolff, 2025.
2. Himalayan cities such as Thimpu, Bhutan are seeing increasing built up area, with repercussions for green cover and exposure to hazards. Chandni Singh, 2018.
3. Waste disposal in a Kampung (urban village) near Jakarta (Indonesia). Erich Wolff, 2019.
4. Waste disposal and accumulation along the Mutha riverbank in Pune (India). Prathijna Poonacha.
5. Informal settlements in Bangkok (Thailand) are prone to flooding. Erich Wolff, 2022.
6. Informal settlement area in Bhopal (India) face heightened heat risk. Chandni Singh, 2025.
7. Constructed wetlands are used in many cities in Southeast Asia, including Putrajaya (Malaysia) and Singapore. Erich Wolff, 2025.
8. Learning spaces at the Hive, campus of the Nanyang Technological University (Singapore). Erich Wolff, 2024.
9. Asia team authors in Miami during one of NATURA Global Roadmap for Urban Nature-based Solutions cross-regional workshops. NATURA, 2023.
10. Urban nature in the Zhejiang Province, China. Yuan Wang.
11. Benchakitti Forest Park in Bangkok, Thailand. Erich Wolff, 2023.
12. Satoyama landscape in the Minamisanriku-Town, Miyagi prefecture, Japan. Yasuo Takahashi.
13. Rice paddies produce food and reduce the risk of floods in the periphery of Chiang Mai (Thailand). Erich Wolff, 2024.
14. Green cover in Mandalay, Myanmar's second largest city. Chandni Singh, 2017.
15. In cities such as Ho Chi Minh City (Vietnam), cultural and spiritual connections with nature have been actively promoted. Erich Wolff, 2025.
16. Poster for the 'Nature and Climate Nexus Webinar Series: Promoting Nature-based Solutions for Sustainable Infrastructure in Asia and the Pacific', hosted by ADB. ADB, 2022.
17. Poster of the 4th webinar under the "Sustainable Urban Development – Smart Cities -II" series organized by National Institute for Urban Affairs (NIUA), India. India Forum For Nature Based Solutions, 2023.
18. Gentrification has been observed near the Chulalongkorn University Centenary Park in Bangkok (Thailand). Erich Wolff, 2023.
19. Green roof in a private commercial development in Singapore. Erich Wolff, 2025.
20. Solar panels installed with their share being used for growing edible plants at ward-level office of the Pune Municipal Corporation, Pune, India. Maitreyi Koduganti, 2025.
21. Urban green parks, such as in Yangon city, Myanmar, serve as a critical green space in dense cities across Asia. Chandni Singh, 2017.
22. River daylighting in Cheonggyecheon, Seoul (South Korea). Seunghee Han, n.d.
23. Terrace Garden at Little Flower School, Adambakkam, Chennai, India. Chennai Resilience Centre, n.d.
24. ICDS terrace garden in an Anganwadi in Nungambakkam, 2024. Chennai Resilience Centre, n.d.
25. Construction of wetlands for wastewater treatment in informal settlements within the RISE programme (Indonesia). Erich Wolff, 2019.
26. Locally-grown plants used in a constructed wetland to treat wastewater within the RISE programme, Makassar (Indonesia). Erich Wolff, 2019.
27. The mangrove restoration project ProCoast is protecting waterfront areas in Iloilo using a participatory approach (the Philippines). Arnel Murga, 2025.
28. Maha Bandula Park in Yangon (Myanmar) provides shade in the centre of the city. Chandni Singh, 2017.
29. Urban greening improves liveability but has also been used to increase property value in commercial developments. Bonifacio High Street, Taguig, Metro Manila (The Philippines). Erich Wolff, 2025.
30. Waterways are a key part of daily life. The River Aerawaddy in Myanmar is used for transport, leisure, and central to local religious life. Chandni Singh, 2018.
31. Model used for community engagement in the RISE program (Indonesia), developed by Dasha Moschonas. Erich Wolff, 2019.
32. Participatory mapping exercise to plan nature-based solutions in Rangsit (Thailand). Erich Wolff, 2023.
33. Asih Busiadi, GCoM Asia Team Leader and Thunpicha Pinniam, GCoM Asia Project Assistant. DAI.com for the European Union, 2022.

6 Appendices

A Details of the Methods Applied (Task 1 and Task 2)

This section outlines the common protocol used across all regional assessments of the *Global Roadmap for Urban Nature-based Solutions*, along with region-specific methodological choices. Together, these methods synthesize the academic literature on urban NbS and assess geographical, bibliometric, and thematic trends. The overall process is summarized in Figure 12.

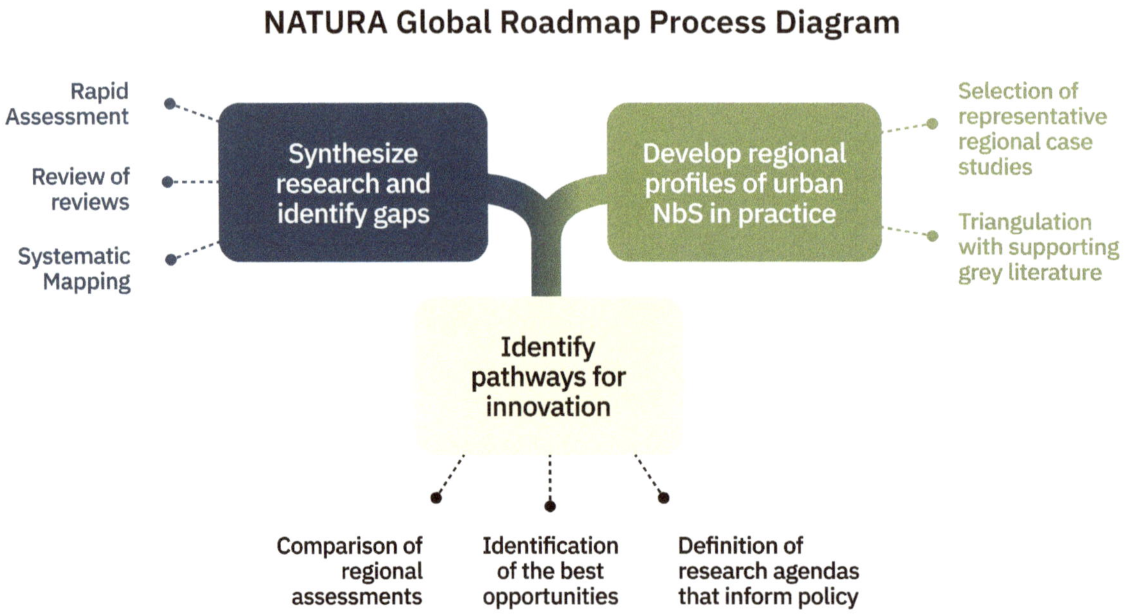

Figure 12: Overall process of the Global Roadmap for Urban Nature-based Solutions

Task 1 | Literature Review

Rapid Assessment

Task 1.1. provides a high-level mapping of the academic landscape on nature-based solutions and related concepts. A Scopus search (36,123 papers) was conducted using a comprehensive set of NbS-related terms, restricted to English and urban contexts. Automated scripts in RStudio identified both regional focus (based on country and city keywords) and conceptual categories (based on NbS-related terms).

Validation checks showed 97% accuracy for region matching and 95.9% for concept matching. The analysis examined temporal publication trends, geographic distributions, and terminological overlaps to position NbS within the broader environmental literature and inform subsequent review and mapping tasks.

The terms explored included:

- blue infrastructure,
- blue-green infrastructure,
- disaster-risk reduction,
- eco-tourism,
- ecological engineering,
- ecological infrastructure,
- ecological restoration,
- ecosystem services,
- ecosystem-based adaptation,
- ecosystem-based approaches,
- ecosystem-based management,
- ecosystem restoration,
- green infrastructure,
- green lungs,
- green space management,
- green development,
- greenbelt,
- landscape ecology approach,
- linear parks,
- natural infrastructure,
- resilient infrastructure,
- sponge city,
- sustainable land management,
- sustainable urban drainage,
- traditional ecological knowledge ,
- urban forestry,
- urban greening,
- urban resilience, and
- water-sensitive urban design.

Review of Reviews

A systematic review of reviews was undertaken to synthesize how urban NbS are conceptualized, reviewed, and discussed in the academic literature.

A Scopus search ("nature-based solution*" AND "review" AND ("urban" OR "city" OR "cities")) identified 450 papers, of which 203 met inclusion criteria after screening. Eligible papers were limited to English-language, urban-focused, and a range of review types (systematic, scoping, bibliometric, targeted, narrative, integrative, comparative, qualitative, critical, rapid, structured, and scientometric reviews), as well as systematic maps, quantitative systematic literature reviews, meta-analyses, bibliographic reviews, thematic analyses, and network analyses).

Each paper was fully screened and manually coded for:

- NbS concepts referenced (e.g., GI, ES, EbA)
- Type and scope of review
- Geographic and thematic trends, biases, and research gaps
- Reported challenges (knowledge system, ecological, social, governance)
- Opportunities and recommendations for future research and practice

This step generated a structured understanding of the key directions and limitations shaping the global NbS knowledge base.

Systematic Mapping

Task 1.3. systematically mapped empirical, place-based studies on urban NbS using searches in Scopus and Web of Science. Search strings combined variants of "nature-based solution(s)" with urban terms (city, cities, urban, town, metropolitan, peri-urban) in titles, abstracts, or author keywords. Results were limited to English-language journal articles published up until the end of 2023. After merging databases and removing duplicates and records without abstracts, 1,436 papers remained for screening; 1,031 papers met the criteria and were included in the final dataset.

Exclusion criteria - papers were included unless they:

- did not contain "nature-based solution(s)" in title, abstract, or keywords
- were not urban in focus
- were not place-based (e.g. conceptual only, no specific study location)
- contained more than ten case studies across multiple world regions (most of these papers were assessed in task 1.2)

Each paper was screened on title and abstract by two reviewers ("yes", "no", or "maybe"), with conflicts resolved through discussion and, for "maybe" cases, full-text checks.

For the retained 1,031 papers, analyses combined manual and automated methods:

- **Manual geographic tagging:** Using information provided in titles, abstracts, and full texts, the team manually assigned each paper to a region, country, city/town, and district/neighbourhood where reported. Papers could receive multiple geographic tags if they covered more than one location.

- **Bibliometric analysis**: Using the bibliometrix package in R, we examined:
    - publication trends over time by region
    - the most frequent journals publishing on urban NbS
    - the most cited papers per region, and
    - the relationship between author affiliation regions and study locations (e.g. proportion of in-country vs external researchers).

These indicators were used to characterise patterns and asymmetries in global research production on urban NbS.

- **Co-occurrence mapping:** Vosviewer was used to generate keyword co-occurrence maps based on author-provided keywords. Papers were split into regional subsets according to the manual tagging, and the minimum keyword occurrence threshold was adjusted to reflect the number of papers in each region. The maps were used to identify dominant thematic clusters and co-occurring concepts across and within regions.

- **Text mining:** Text mining in R was guided by a predefined keyword codebook developed iteratively using expert input, case study data (Task 2), and reference frameworks (e.g. IUCN, NetworkNature). The codebook classified keywords under two main categories — ecosystem types (green, blue, brown, etc.) and societal challenges (climate, ecosystem health, wellbeing, inclusion, etc.) — and was refined through multiple validation rounds until achieving an average coding accuracy of 94.9%.

The resulting dataset provided the empirical foundation for visualizing geographic and thematic research patterns, highlighting where and how NbS are studied across regions.

Task 2 | Case Study Selection

Initially, we compiled a comprehensive list of types of NbS and ecosystems in Asia. An in-depth examination of the literature revealed that urban NbS encompasses a diverse array of strategies and interventions, including green spaces, blue and green infrastructure, and integrated water management approaches. We categorized these solutions into 13 themes (reforestation, forest conservation, agricultural best management practices urban periphery, urban agriculture in central areas, urban wildlife and biodiversity, urban greening including green roof or wall (building scale), urban greening including wetland park, coral restoration, marine ecosystems, mangrove restoration, river re-naturalization, river restoration and wetland conservation) to facilitate the identification of effective case studies from various regions and cities and ensure different types of NbS interventions were covered. Recognizing the diversity of case studies in the region, we established five criteria to guide our classification and analysis (outlined in figure 13).

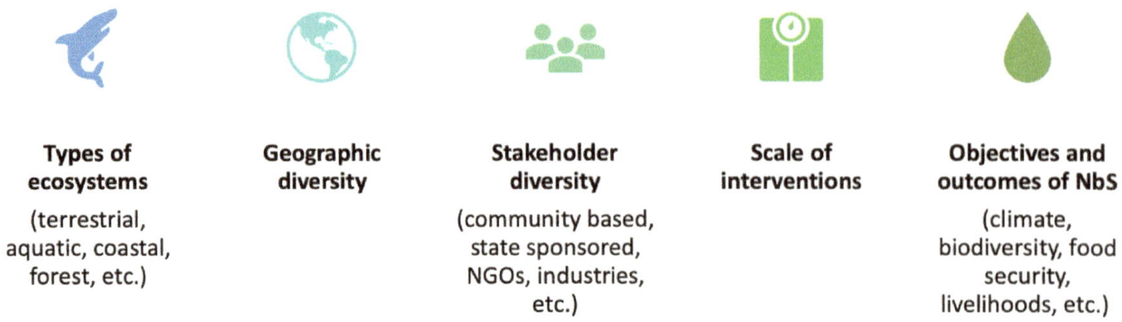

Figure 13: Asia case study selection criteria

Shortlisting Case Studies

For the process of shortlisting case studies, various methods were considered, including a scoring method based on selected criteria, the selection of cases based on available information, and possibly focusing on cases that were less documented. However, these methods did not yield expected results as they introduced biases based on the weights selected for each criterion and were not representative of the diversity and trends as observed in NbS in the region.

Drawing on the lessons learnt from the earlier attempts, a more heuristic approach emerged which considered different aspects of each of these methods. Utilizing each of NbS themes as the primary filter, we systematically shortlisted corresponding case studies under each NbS category. Additional filters such as country, type of stakeholder, and scale were employed to further refine the shortlist. This comprehensive and multi-dimensional approach enabled us to arrive at a more robust and meaningful set of case studies that are more representative of the rich diversity of the region and of its growing database of NbS.

The process resulted in the first set of 27 case studies from 16 countries, which offers entries that are more representative of the regions and the different types of NbS implemented.

Forming the Regional Expert Team

As the team worked on selecting case studies, they simultaneously identified experts with in-depth knowledge of the respective regions and cases. These experts were identified either from within their own network or through extended professional connections. Once suitable experts were identified, the regional leads formally invited them to contribute to the case study write-ups. Throughout the process, the regional leads and fellows engaged in continuous collaboration with the experts, ensuring their insights and expertise were effectively incorporated. This co-development approach involved regular interactions and discussions, with each case study undergoing two rigorous rounds of review before finalization

B Summary of Case Studies

Table 2: Summary of Asia case studies (cases AS_01 to AS_03)

ID (AS_)	Project name	Country	Societal problems addressed	Co-benefits	Impact scale	Urban typology	Post-intervention ecosystem category	Post-intervention ecosystem subcategory	NbS typology	NbS techniques (general)	NbS techniques (specific)
01	**Benjakitti Forest Park**	Thailand	Social wellbeing; Physical and mental health	Flood mitigation; Increased biodiversity and their functions; Recreational benefits	Neighbourhood/ community	Urban core	Green; Blue	Green spaces (gen.); Rivers & waterways; Forests & trees; Stormwater infrastructure	Restoration; Constructed & novel ecosystems	Revegetation & planting (general); Species conservation & monitoring; Forestation & urban tree strategies; Terrestrial rehabilitation & site remediation; Freshwater ecosystem creation, restoration & management; Wetland conservation, restoration & construction	Vegetation planting; Native & indigenous species planting; Tree planting (outside of forests); Constructed wetlands; Constructed ponds; Post-industrial and contaminated site rehabilitation
02	**Bishan Ang Mo Kio Park**	Singapore	Physical and mental health; Flooding	Recreational benefits; Water quality; Reinforcement of socio-ecological knowledge; Increased biodiversity and their functions	Neighbourhood/ community	Urban core	Blue-green; Blue; Green	Wetlands; Rivers & waterways; Green spaces (gen.)	Restoration; Constructed & novel ecosystems	Erosion risk reduction & soil remediation; Wastewater treatment; Freshwater ecosystem creation, restoration & management; Wetland conservation, restoration & construction	River restoration (gen.); Constructed riparian buffer zones; Stabilization; Constructed wetlands; Bioswales, biofiltration & rain gardens
03	**Chongqing Urban Regeneration**	China	Flooding; Nature connection	Recreational benefits; Place regeneration & urban revitalization; Increased biodiversity and their functions; Water quality	City	Urban core	Green; Blue-green	Green spaces (gen.); Wetlands; Stormwater infrastructure	Restoration; Constructed & novel ecosystems	Urban agricultural practices; Stormwater absorption, retention & treatment; Recreational & private green space management; Urban green infrastructure construction; Erosion risk reduction & soil remediation; Revegetation & planting (general); Forestation & urban tree strategies; Terrestrial rehabilitation & site remediation	Wetland restoration; Retention ponds & basins; Recreational green space management (gen.); Green alleys & corridors

ID (AS_)	Project name	Country	Primary taxonomy	Domestic government role	Foreign government role	Non-governmental actors role	Private sector role	Academic actors role	Community role	Indigenous group	Institutional instrument
01	**Benjakitti Forest Park**	Thailand	Ecological or natural infrastructure	Project lead; Primary funder	No role	No role	Specialized knowledge	No role	Unknown	0	Unknown
02	**Bishan Ang Mo Kio Park**	Singapore	Green infrastructure (GI)	Project lead; Primary funder	No role	No role	Unknown	No role	No role	0	Policy
03	**Chongqing Urban Regeneration**	China	Green infrastructure (GI)	Project lead; Primary funder	No role	Support funder; Specialized knowledge	Unknown	No role	Implementation partner	0	Plan (e.g., Master plan)

ID (AS_)	Project name	Country	Project timeline			Status	Budget	Implementation barriers	Implementation enablers
			Start	End	Duration (yrs)				
01	**Benjakitti Forest Park**	Thailand	2019		6	Ongoing	10,000,000 - 50,000,000	Poor ecological health	Good coordination or collaboration; Innovative and adapted design; Technical expertise
02	**Bishan Ang Mo Kio Park**	Singapore	2009	2012	3	Complete	5,000,000 - 10,000,000	Unknown	Good coordination or collaboration; Education, awareness, and advocacy
03	**Chongqing Urban Regeneration**	China	2018		7	Ongoing	100,000,000 - 1,000,000,000	Unknown	Funding available; Innovative and adapted design; Technical expertise

Table 3: Summary of Asia case studies (cases AS_04 to AS_08)

ID (AS_)	Project name	Country	Societal problems addressed	Co-benefits	Impact scale	Urban typology	Post-intervention ecosystem category	Post-intervention ecosystem subcategory	NbS typology	NbS techniques (general)	NbS techniques (specific)
04	**Palm Geotextile To Prevent Erosion**	Vietnam	Land degradation; Soil quality	Green jobs and economic benefits; Water quality	Greater urban region	Outside the city with urban impacts	Blue-green; Blue	Riparian areas; Rivers & waterways	Constructed & novel ecosystems	Erosion risk reduction & soil remediation; Biomaterials & biogas production	Erosion risk reduction; Biodegradable woven mats
05	**Pingshan River Water Treatment Wetlands**	China	Water quality	Increased biodiversity and their functions; Flood mitigation; Recreational benefits; Green jobs and economic benefits	Greater urban region	Metropolitan	Blue	Wetlands; Rivers & waterways	Constructed & novel ecosystems	Wetland conservation, restoration & construction; Wastewater treatment	Constructed wetlands; Water treatment (gen.)
06	**Shenzhen Mangrove Restoration**	China	Land degradation; Nature connection; Biodiversity loss	Urban cooling; Carbon sequestration; Water quality	Greater urban region	Peri-urban	Blue-green; Blue	Mangroves; Lakes & ponds; Coastal (gen.)	Protection & conservation; Restoration	Mangrove conservation & restoration; Revegetation & planting (general); Freshwater ecosystem creation, restoration & management	Mangrove restoration; Mangrove conservation; Vegetation planting; Pond restoration
07	**Urchinomics**	Japan	Biodiversity loss	Carbon sequestration; Green jobs and economic benefits	Transboundary (subnational)	Metropolitan	Blue	Productive blue spaces	Restoration	Urban agricultural practices	Aquaculture
08	**Gangxia 1980 Project: Green Roof For Youth'S Mental Health**	China	Social wellbeing; Flooding	Place regeneration & urban revitalization; Stronger human-nature connections; Recreational benefits	Building	Urban core	Green	Roofs & walls; Agriculture & food spaces	Constructed & novel ecosystems	Urban green infrastructure construction; Stormwater absorption, retention & treatment	Green roofs; Rainwater harvesting (gen.)

ID (AS_)	Project name	Country	Primary taxonomy	Domestic government role	Foreign government role	Non-governmental actors role	Private sector role	Academic actors role	Community role	Indigenous group	Institutional instrument
04	**Palm Geotextile To Prevent Erosion**	Vietnam	Biogeotextiles	Unknown	Primary funder	Unknown	Unknown	Project lead	Participant	0	Unknown
05	**Pingshan River Water Treatment Wetlands**	China	Nature-based solutions	Project lead; Primary funder	No role	No role	No role	Specialized knowledge	No role	0	Unknown
06	**Shenzhen Mangrove Restoration**	China	Nature-based solutions	Project lead	No role	Primary funder	Unknown	Unknown	Unknown	0	International agreements
07	**Urchinomics**	Japan	Nature-based solutions	No role	No role	No role	Project lead; Primary funder	Research media and dissemination	Implementation partner	0	Unknown
08	**Gangxia 1980 Project: Green Roof For Youth'S Mental Health**	China	Green infrastructure (GI)	Participant	No role	Project lead; Primary funder	Implementation partner; Specialized knowledge	Unknown	Participant	0	Policy

ID (AS_)	Project name	Country	Project timeline			Status	Budget	Implementation barriers	Implementation enablers
			Start	End	Duration (yrs)				
04	**Palm Geotextile To Prevent Erosion**	Vietnam	2005	2008	3	Complete	1,000,000 - 5,000,000	Resistance to change & cultural barriers	Equity-centered; Stakeholder engagement; Co-production approaches in management
05	**Pingshan River Water Treatment Wetlands**	China	2018		7	Ongoing	1,000,000 - 5,000,000	Maintenance challenges	Unknown
06	**Shenzhen Mangrove Restoration**	China	1988		37	Ongoing	10,000,000 - 50,000,000	Poor ecological health	Funding available
07	**Urchinomics**	Japan	2016		9	Ongoing	Not applicable	Poor ecological health	Low costs; Economic viability
08	**Gangxia 1980 Project: Green Roof For Youth'S Mental Health**	China	2017	2020	3	Complete	Unknown	Resistance to change & cultural barriers; Implementation delays	Enabling policies and regulations; Innovative and adapted design; Technical expertise

Table 4: Summary of Asia case studies (cases AS_09 to AS_13)

ID (AS_)	Project name	Country	Societal problems addressed	Co-benefits	Impact scale	Urban typology	Post-intervention ecosystem category	Post-intervention ecosystem subcategory	NbS typology	NbS techniques (general)	NbS techniques (specific)
09	**Shijiu Yang Wetland Park For Water Treatment**	China	Water quality; Water security	Human health and well-being	Trans-boundary (subnational)	Peri-urban	Blue-green; Green	Wetlands; Corridors; Stormwater infrastructure	Constructed & novel ecosystems	Wastewater treatment; Revegetation & planting (general); Stormwater absorption, retention & treatment; Wetland conservation, restoration & construction	Water treatment (gen.); Bioswales, biofiltration & rain gardens; Constructed wetlands; Marine vegetation & wildlife management
10	**The Floating Gardens Of Inle Lake**	Myanmar	Water quality; Food security; Economic vunerability	Increased biodiversity and their functions; Reinforcement of socio-ecological knowledge	Greater urban region	Peri-urban	Blue; Blue-green	Lakes & ponds; Wetlands	Restoration	Wastewater treatment	Floating gardens & islands
11	**Procoast'S Sustainable Coastal Protection**	Philippines	Sea level rise; Storms; Social inclusion	Increased biodiversity and their functions; Reinforcement of socio-ecological knowledge	City	Peri-urban	Blue; Blue-green	Mangroves; Productive blue spaces	Protection & conservation; Restoration	Mangrove conservation & restoration; Urban agricultural practices	Mangrove conservation; Mangrove restoration; Aquaculture
12	**Conservation Of Subak Systems**	Indonesia	Food security; Economic vunerability; Soil quality	Increased biodiversity and their functions; Reinforcement of socio-ecological knowledge; Water security	Greater urban region	Peri-urban	Blue-green; Green	Wetlands; Agriculture & food spaces	Protection & conservation; Restoration	Urban agricultural practices	Subak irrigation
13	**Urban Agriculture At Solar-Powered Kochi International Airport**	India	Food security; Social inclusion	Green jobs and economic benefits; Carbon sequestration; Urban cooling	Building	Urban core	Green	Agriculture & food spaces	Constructed & novel ecosystems	Urban agricultural practices	Urban farming (gen.)

ID (AS_)	Project name	Country	Primary taxonomy	Domestic government role	Foreign government role	Non-governmental actors role	Private sector role	Academic actors role	Community role	Indigenous group	Institutional instrument
09	**Shijiu Yang Wetland Park For Water Treatment**	China	Ecological restoration	Project lead; Primary funder	No role	No role	Implementation partner	Research media and dissemination; Specialized knowledge	Participant	0	Plan (e.g., Master plan)
10	**The Floating Gardens Of Inle Lake**	Myanmar	Sustainable land management	Participant	Support funder; Specialized knowledge	Support funder; Implementation partner	No role	No role	Project lead; Maintenance and/or stewardship	1	Unknown
11	**Procoast'S Sustainable Coastal Protection**	Philippines	Sustainable land management	Project lead	Primary funder; Project lead	Specialized knowledge	No role	Research media and dissemination	Implementation partner	1	Policy
12	**Conservation Of Subak Systems**	Indonesia	Sustainable land management	Support funder	No role	Support funder	No role	No role	Project lead; Maintenance and/or stewardship; Specialized knowledge	1	Unknown
13	**Urban Agriculture At Solar-Powered Kochi International Airport**	India	Nature-based solutions	Project lead	No role	No role	Project lead; Primary funder	No role	No role	0	Urban strategy

ID (AS_)	Project name	Country	Project timeline			Status	Budget	Implementation barriers	Implementation enablers
			Start	End	Duration (yrs)				
09	**Shijiu Yang Wetland Park For Water Treatment**	China	2005	2009	4	Complete	10,000,000 - 50,000,000	Poor ecological health	Funding available; Mel frameworks; Knowledge production and sharing; Education, awareness, and advocacy
10	**The Floating Gardens Of Inle Lake**	Myanmar				Ongoing	Not applicable	Unknown	Environmental stewardship
11	Procoast'S Sustainable Coastal Protection	Philippines	2018	2021	3	Complete	Unknown	Unknown	Unknown
12	**Conservation Of Subak Systems**	Indonesia				Ongoing	Not applicable	Values and/or priorities conflicts	Cultural integration
13	**Urban Agriculture At Solar-Powered Kochi International Airport**	India	2013	2021	8	Complete	250,000 - 500,000	Design adaptation and skills gaps; Maintenance challenges	Strong leadership

Table 5: Summary of Asia case studies (cases AS_14 to AS_18)

ID (AS_)	Project name	Country	Societal problems addressed	Co-benefits	Impact scale	Urban typology	Post-intervention ecosystem category	Post-intervention ecosystem subcategory	NbS typology	NbS techniques (general)	NbS techniques (specific)
14	**Puey Ungphakorn Centenary Hall Green Roof**	Thailand	Flooding; Heat	Reinforcement of socio-ecological knowledge; Community engagement & social cohesion; Food security	Building	Peri-urban	Green	Roofs & walls; Green spaces (gen.)	Constructed & novel ecosystems	Urban green infrastructure construction; Urban agricultural practices; Stormwater absorption, retention & treatment"	Green roofs; Water sensitive urban design; Vegetable gardening
15	**Putrajaya Wetlands Park**	Malaysia	Flooding; Water quality; Nature connection	Increased biodiversity and their functions; Recreational benefits	Neighbourhood/community	Urban core	Blue-green; Green	Riparian areas; Green spaces (gen.); Wetlands	Restoration; Constructed & novel ecosystems	Recreational & private green space management; Wetland conservation, restoration & construction	Recreational green space management (gen.); Constructed wetlands
16	**Fukuoka Hall Green Roof**	Japan	Physical and mental health; Social well-being; Social inclusion	Recreational benefits; Increased biodiversity and their functions; Flood mitigation; Urban cooling	Building	Urban core	Green	Roofs & walls	Constructed & novel ecosystems	Urban green infrastructure construction; Species conservation & monitoring	Green roofs; Vertical gardens (non-agricultural); Native & indigenous species planting
17	**From Farm To Classroom: Bengaluru'S Ragi Project**	India	Nature connection; Land degradation	Reinforcement of socio-ecological knowledge; Community engagement & social cohesion; Food security	Site	Urban core	Green	Agriculture & food spaces	Constructed & novel ecosystems; Restoration	Urban agricultural practices	Urban farming (gen.); Schoolyard greening
18	**Making Space For Fauna**	India	Land degradation; Water quality; Biodiversity loss; Heat	Food security; Reinforcement of socio-ecological knowledge; Green jobs and economic benefits	Site	Urban core	Green; Blue; Blue-green	Agriculture & food spaces; Lakes & ponds; Wetlands	Constructed & novel ecosystems; Restoration	Species conservation & monitoring; Wastewater treatment; Urban agricultural practices; Freshwater ecosystem creation, restoration & management	Bird, insect & mammal shelters; Floating gardens & islands; Constructed ponds; Urban farming (gen.)

ID (AS_)	Project name	Country	Primary taxonomy	Domestic government role	Foreign government role	Non-governmental actors role	Private sector role	Academic actors role	Community role	Indigenous group	Institutional instrument
14	**Puey Ungphakorn Centenary Hall Green Roof**	Thailand	Water sensitive urban drainage (WSUD)	No role	No role	No role	No role	Project lead; Primary funder; Maintenance and/or stewardship	Maintenance and/or stewardship	0	Urban strategy
15	**Putrajaya Wetlands Park**	Malaysia	Blue infrastructure	Project lead; Primary funder	No role	No role	Unknown	Unknown	Participant	0	Plan (e.g., Master plan); Policy
16	**Fukuoka Hall Green Roof**	Japan	Green roof	Project lead; Primary funder	No role	No role	No role	No role	No role	0	Urban strategy
17	**From Farm To Classroom: Bengaluru'S Ragi Project**	India	Farm-based learning	No role	No role	Implementation partner	Support funder	Project lead; Specialized knowledge	Implementation partner; Maintenance and/or stewardship	0	Program
18	**Making Space For Fauna**	India	Nature-based solutions	Support funder	Implementation partner	Support funder	Unknown	Project lead; Specialized knowledge	Participant	1	Urban strategy

ID (AS_)	Project name	Country	Project timeline			Status	Budget	Implementation barriers	Implementation enablers
			Start	End	Duration (yrs)				
14	**Puey Ungphakorn Centenary Hall Green Roof**	Thailand	2016	2019	3	Complete	10,000,000 - 50,000,000	Unknown	Unknown
15	**Putrajaya Wetlands Park**	Malaysia	1997	2025	28	Ongoing	Unknown	High costs & economic risks	Institutional capacity (e.g., Decentralized approaches); Enabling policies and regulations
16	**Fukuoka Hall Green Roof**	Japan	1995	1995	0	Complete	1,000,000 - 5,000,000	Resistance to change & cultural barriers	Innovative and adapted design; Technical expertise
17	**From Farm To Classroom: Bengaluru'S Ragi Project**	India	2017	2019	2	Complete	Unknown	Design adaptation and skills gaps; Poor ecological health	Education, awareness, and advocacy; Mel frameworks; Knowledge production and sharing
18	**Making Space For Fauna**	India	2021		4	Ongoing	Unknown	Maintenance challenges	Equity-centered; Stakeholder engagement; Co-production approaches in management

Table 6: Summary of Asia case studies (cases AS_19 to AS_22)

ID (AS_)	Project name	Country	Societal problems addressed	Co-benefits	Impact scale	Urban typology	Post-intervention ecosystem category	Post-intervention ecosystem subcategory	NbS typology	NbS techniques (general)	NbS techniques (specific)
19	**Tashkent Green City**	Uzbekistan	Heat; Air quality; Nature connection	Flood mitigation; Recreational benefits; Human health and well-being	City	Metropolitan	Green	Forests & trees; Green spaces (gen.)	Restoration	Forestation & urban tree strategies; Recreational & private green space management	Tree planting (outside of forests); Recreational green space management (gen.)
20	**Youth-Led Rehabilitation Of Coral Reefs In The Kayaker Coral Cove**	Sri Lanka	Sea level rise; Biodiversity loss	Reinforcement of socio-ecological knowledge; Stronger human-nature connections	Site	Outside the city with urban impacts	Blue	Reefs	Restoration; Protection & conservation	Species conservation & monitoring; Intervention to coastal and marine habitats; Forestation & urban tree strategies	Reef restoration & management; Solid waste management (gen.)
21	**Mangrove Protection In The Indus Delta**	Pakistan	Biodiversity loss; Storms; Land degradation	Green jobs and economic benefits; Carbon sequestration	Site	Outside the city with urban impacts	Blue-green	Mangroves; Riparian areas	Restoration	Mangrove conservation & restoration; Species conservation & monitoring; Forestation & urban tree strategies	Mangrove restoration; Reforestation; Species nurseries; Afforestation
22	**East Kolkata Wetlands**	India	Flooding; Water quality; Heat; Biodiversity loss	Reduced infrastructure cost; Food security; Carbon sequestration; Air quality; Recreational benefits	Greater urban region	Outside the city with urban impacts	Blue-green	Wetlands	Protection & conservation; Restoration	Wetland conservation, restoration & construction	Wetland restoration; Wetland conservation

ID (AS_)	Project name	Country	Primary taxonomy	Domestic government role	Foreign government role	Non-governmental actors role	Private sector role	Academic actors role	Community role	Indigenous group	Institutional instrument
19	**Tashkent Green City**	Uzbekistan	Nature-based solutions	Project lead	No role	Primary funder; Implementation partner; Specialized knowledge	No role	No role	Implementation partner; Maintenance and/or stewardship	0	Program; Urban strategy
20	**Youth-Led Rehabilitation Of Coral Reefs In The Kayaker Coral Cove**	Sri Lanka	Nature-based solutions	Participant	No role	Specialized knowledge; Support funder; Participant	Support funder	Research media and dissemination; Specialized knowledge; Implementation partner	Project lead; Maintenance and/or stewardship; Support funder	0	Program; Policy
21	**Mangrove Protection In The Indus Delta**	Pakistan	Nature-based solutions	Project lead	No role	Project lead; Primary funder; Specialized knowledge	No role	Unknown	Implementation partner; Maintenance and/or stewardship	0	Policy
22	**East Kolkata Wetlands**	India	Ecological or natural infrastructure	Project lead; Primary funder	No role	Research media and dissemination; Specialized knowledge	Implementation partner	Research media and dissemination	Implementation partner	0	Law

ID (AS_)	Project name	Country	Project timeline			Status	Budget	Implementation barriers	Implementation enablers
			Start	End	Duration (yrs)				
19	**Tashkent Green City**	Uzbekistan	2019		6	Ongoing	Unknown	Unknown	Enabling policies and regulations
20	**Youth-Led Rehabilitation Of Coral Reefs In The Kayaker Coral Cove**	Sri Lanka	2022		3	Ongoing	< 100,000	Funding gaps; External disruptions	Equity-centered; Stakeholder engagement; Co-production approaches in management; Good coordination or collaboration; Enabling policies and regulations; Low costs; Economic viability
21	**Mangrove Protection In The Indus Delta**	Pakistan	2007		18	Ongoing	Unknown	Maintenance challenges	Funding available; Education, awareness, and advocacy
22	**East Kolkata Wetlands**	India	1992		33	Ongoing	10,000,000 - 50,000,000	Values and/or priorities conflicts; Resistance to change & cultural barriers; Monitoring, evaluation and learning gaps	Institutional capacity (e.g., Decentralized approaches)

Table 7: Summary of Asia case studies (cases AS_23 to AS_25)

ID (AS_)	Project name	Country	Societal problems addressed	Co-benefits	Impact scale	Urban typology	Post-intervention ecosystem category	Post-intervention ecosystem subcategory	NbS typology	NbS techniques (general)	NbS techniques (specific)
23	**Improving Degraded Riverine Areas In Rohingya Camps Of Ukhiya, Cox'S Bazar**	Bangladesh	Water security; Water quality; Social inclusion	Food security; Increased biodiversity and their functions; Flooding	Neighbourhood/community	Outside the city with urban impacts	Blue-green; Blue; Green	Rivers & waterways; Riparian areas; Forests & trees; Wetlands	Restoration; Protection & conservation; Constructed & novel ecosystems	Forestation & urban tree strategies; Wastewater treatment; Species conservation & monitoring; Revegetation & planting (general); Freshwater ecosystem creation, restoration & management; Stormwater absorption, retention & treatment; Wetland conservation, restoration & construction	River restoration (gen.); Afforestation; Reforestation; Silt traps; Wastewater management (gen.); Native & indigenous species planting; Constructed wetlands; Vegetation planting; Species nurseries
24	**Female-Led Turtle Conservation, Reef Restoration, And Sustainable Fishing**	India	Economic vunerability; Biodiversity loss	Carbon sequestration; Community engagement & social cohesion; Stronger human-nature connections	Building	Outside the city with urban impacts	Blue	Rivers & waterways	Protection & conservation	Intervention to coastal and marine habitats; Species conservation & monitoring	Marine vegetation & wildlife management; Artificial reef construction; Species conservation (gen.)
25	**Oyster Breakwater Reefs Support Climate Adaptation' In Bangladesh**	Bangladesh	Biodiversity loss; Sea level rise; Land degradation	Water quality	Site	Outside the city with urban impacts	Blue	Reefs	Restoration	Intervention to coastal and marine habitats	Artificial reef construction; Marine vegetation & wildlife management

ID (AS_)	Project name	Country	Primary taxonomy	Domestic government role	Foreign government role	Non-govern-mental actors role	Private sector role	Academic actors role	Community role	Indig-enous group	Institutional instrument
23	**Improving Degraded Riverine Areas In Rohingya Camps Of Ukhiya, Cox'S Bazar**	Bangla-desh	Ecological resto-ration	Participant	Implementa-tion partner	Project lead; Primary funder	No role	Specialized knowledge	Maintenance and/or stew-ardship	0	Law
24	**Female-Led Turtle Conserva-tion, Reef Resto-ration, And Sustainable Fishing**	India	Green develop-ment	No role	No role	Project lead; Maintenance and/or steward-ship; Special-ized knowledge	Participant	No role	Project lead; Implementa-tion partner; Maintenance and/or stew-ardship	0	Program
25	**Oyster Breakwa-ter Reefs Support Climate Adaptation' In Bangla-desh**	Bangla-desh	Nature-based solutions	No role	No role	No role	No role	Project lead; Primary funder; Research media and dissemina-tion	Implementa-tion partner; Maintenance and/or stew-ardship	1	Program

ID (AS_)	Project name	Country	Project timeline			Status	Budget	Implementation barriers	Implementation enablers
			Start	End	Duration (yrs)				
23	**Improving Degraded Riverine Areas In Rohingya Camps Of Ukhiya, Cox'S Bazar**	Bangla-desh	2019	2021	2	Complete	250,000 - 500,000	Maintenance challenges; Poor ecolog-ical health	Equity-centered; Stakeholder engage-ment; Co-production approaches in management; Good coordination or collab-oration; Funding available; Environmental stewardship
24	**Female-Led Turtle Conserva-tion, Reef Restoration, And Sustain-able Fishing**	India	1993		32	Ongoing	Unknown	Weak institutional capacity; Mainte-nance challenges	Equity-centered; Stakeholder engage-ment; Co-production approaches in management; Good coordination or collab-oration; Strong leadership; Institutional capacity (e.g., Decentralized approaches); Low costs; Economic viability; Education, awareness, and advocacy
25	**Oyster Breakwa-ter Reefs Support Climate Adaptation' In Bangla-desh**	Bangla-desh	2012	2018	6	Complete	Unknown	Poor ecological health; External disrup-tions	Equity-centered; Stakeholder engage-ment; Co-production approaches in management; Cultural integration; Low costs; Economic viability; Mel frameworks; Knowledge production and sharing

Table 8: Summary of Asia case studies (cases AS_26 to AS_27)

ID (AS_)	Project name	Country	Societal problems addressed	Co-benefits	Impact scale	Urban typology	Post-intervention ecosystem category	Post-intervention ecosystem subcategory	NbS typology	NbS techniques (general)	NbS techniques (specific)
26	**Korea-Mongolia Greenbelt Reforestation Project**	Mongolia, South Korea	Land degradation; Storms; Drought	Carbon sequestration; Stronger human-nature connections; Recreational benefits; Green jobs and economic benefits; Reduced infrastructure cost	Multi-national	Metropolitan	Green	Green spaces (gen.)	Restoration	Forestation & urban tree strategies; Species conservation & monitoring	Tree planting (outside of forests); Species nurseries; Afforestation
27	**Cheonggyecheon Park**	South Korea	Flooding; Physical and mental health; Nature connection	Water quality; Air quality; Increased biodiversity and their functions; Green jobs and economic benefits	Neighbourhood/ community	Urban core	Blue; Blue-green	Rivers & waterways; Stormwater infrastructure	Restoration	Recreational & private green space management; Freshwater ecosystem creation, restoration & management"; Stormwater absorption, retention & treatment"	Water sensitive urban design; Recreational green space construction (gen.); Stream rehabilitation

ID (AS_)	Project name	Country	Primary taxonomy	Domestic government role	Foreign government role	Non-governmental actors role	Private sector role	Academic actors role	Community role	Indigenous group	Institutional instrument
26	**Korea-Mongolia Greenbelt Reforestation Project**	Mongolia, South Korea	Sustainable land management	Project lead	Project lead; Primary funder	No role	No role	Unknown	Participant	0	International agreements
27	**Cheonggyecheon Park**	South Korea	Ecological restoration	Project lead; Primary funder	No role	Participant	Unknown	No role	Participant	0	Plan (e.g., Master plan)

ID (AS_)	Project name	Country	Project timeline			Status	Budget	Implementation barriers	Implementation enablers
			Start	End	Duration (yrs)				
26	**Korea-Mongolia Greenbelt Reforestation Project**	Mongolia, South Korea	2007	2026	19	Ongoing	5,000,000 - 10,000,000	High costs & economic risks; Maintenance challenges	Institutional capacity (e.g., Decentralized approaches); Enabling policies and regulations; Funding available
27	**Cheonggyecheon Park**	South Korea	2002	2005	3	Complete	100,000,000 - 1,000,000,000	Values and/or priorities conflicts	Funding available

C Case Study Write-ups

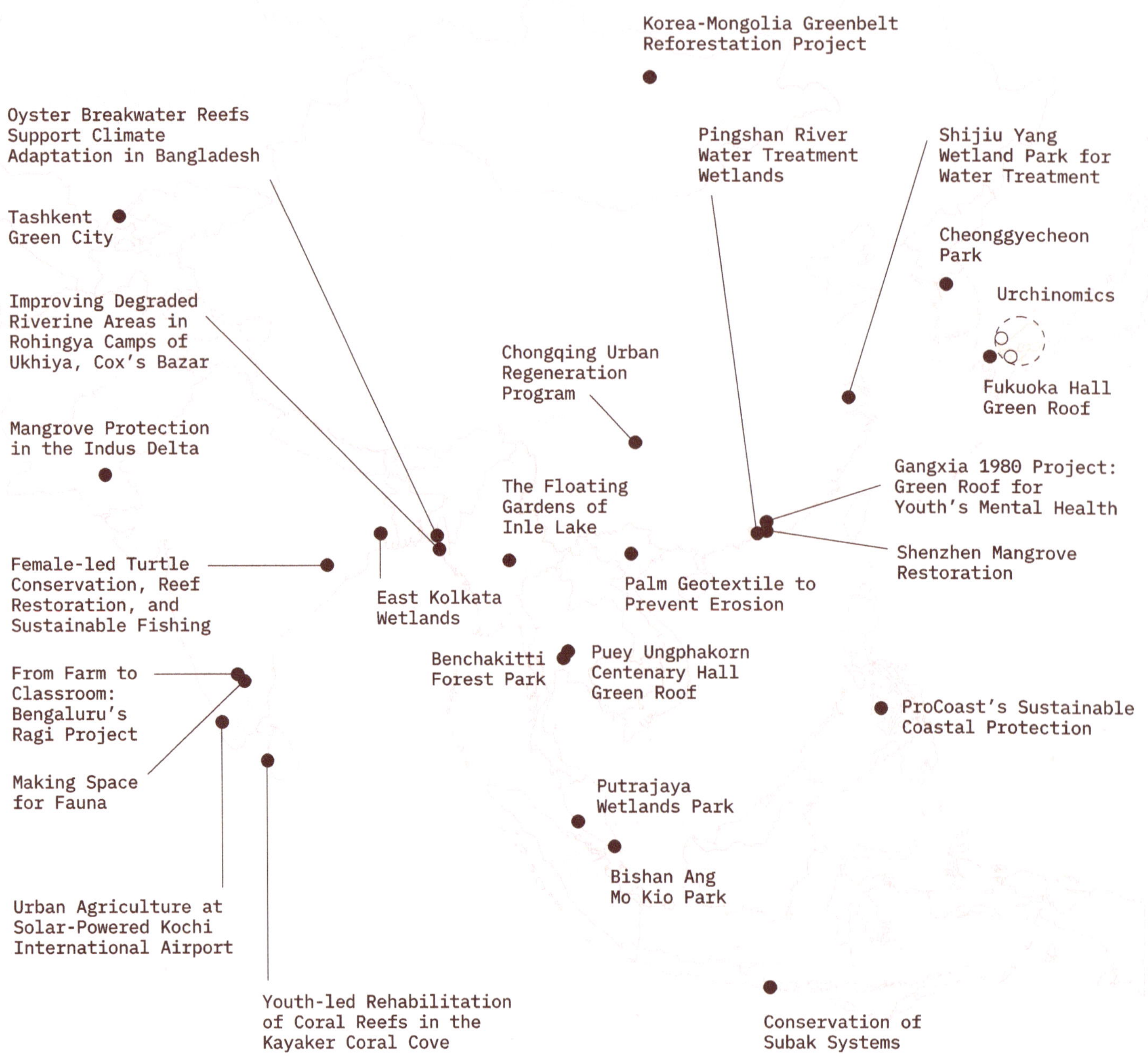

The distribution of urban Nature-based Solutions case studies in Asia.

AS_01 Benchakitti Forest Park

Context Overview

NBS TYPOLOGY Restoration; Constructed & novel ecosystems	ECOSYSTEM TYPE Green spaces (gen.); Rivers & waterways; Forests & trees; Storm-water infrastructure
IMPACT SCALE Neighborhood/Community	URBAN TYPOLOGY Urban core
SOCIAL CHALLENGES ADDRESSED Social wellbeing; Physical and mental health	CO-BENEFITS Flood mitigation; Increased biodiversity and their functions; Recreational benefits

Benchakitti is a park embedded in a dense urban environment in Thailand that illustrates how nature-based solutions can transform brownfields. The park was built on a site previously allocated to industrial activities and plays a key role in restoring urban biodiversity, reducing flood risk and providing cultural and health benefits to society.

The park is located in the central metropolitan area of Bangkok, the capital of Thailand, a low-lying tropical megacity located alongside the Chao Phraya River. The metropolitan region of Bangkok accounts for more than 80% of the total urbanized area of Thailand and has over 10 million inhabitants. Being located in an area with an average temperature of approximately 28 degrees and annual precipitation of around 1,200 mm, the city suffers from a lack of green public spaces as it struggles to adapt to floods and heatwaves.

Introduction to the case study

The park is mostly flat and contains elevated walkways connecting its different parts, bridging over constructed wetlands and lakes. Embedded in the context of the 20-year development plan of Bangkok Metropolis (Huu Loc et al. 2020), the nature-based solutions implemented within Benchakitti were designed as a redevelopment of the previous Benchakitti Water Park, which occupies a land that was formerly used by a tobacco industrial facility.

The redevelopment of the Benchakitti park was motivated by an interest in providing a new public park that complements the historic and more established Lumphini park. Adjacent to some of Bangkok's most dynamic commercial and residential areas, the park is expected to promote sports and cultural activities and provide recreational benefits to the community in a dense urban center. Given that Bangkok is subject to intense flooding (Linh, Ahmed, and Hu Loc 2022), a secondary objective was to reduce floodwater runoff from the area. The area is formally described as a "forest park" in official documents, highlighting that this is a system designed to be managed adaptively with little human intervention after establishment. This concept is aligned with the project goals of providing habitat and biodiversity benefits by redeveloping a formerly industrial facility (Likitswat and Sung 2025).

Nature-based Solutions characteristics

The Benchakitti forest park is composed of a series of pocket forests, green open spaces, sport facilities (including elevated walkways and trails) and an area designated for urban farming (Likitswat and Sung 2025). These facilities are linked to infiltration bioswales and distributed alongside waterways, such as constructed ponds, wetlands and creeks.

The ambitious project being implemented in Benchakitti consists of conserving existing features and actively restoring the ecosystem by renaturalizing previous drainage systems and planting native vegetation. The restoration efforts included adding more than 300 species of native trees and other types of vegetation to the park and redesigning the wetlands to incorporate islands with the objective of increasing oxygen levels in the soil (UNA 2023).

Governance

The project was commissioned and implemented on land owned by the Treasury Department of the Thai government, under the Ministry of Finance as part of ongoing efforts to improve the environmental conditions in Bangkok. Construction works were undertaken by the Royal Thai Army following the design provided by Arsomsilp Community and Environment Architect, which estimated the project to cost about 670 million Baht (USD 20 million). Funded by the Thai Tobacco Authority, the park is currently under the management of the Bangkok Metropolitan Authority (BMA), which is the authority responsible for the maintenance of trees and landscaping in the city (UNA 2023).

Outcome

Due to its significant proportions and ambitious "forest park" concept, the park is being implemented in phases. The design phase began in 2009 and the main recreational facilities, including the green area for events, elevated walkways and constructed wetlands were inaugurated between 2021 and 2022. The construction of the second phase of the project, which will include the renovation of historic buildings within the park's perimeter and the establishment of the urban farming grounds, are still underway. While the park is considered a success by the local government, it has ignited discussions about preferences for different types of nature and aesthetic expectations regarding urban parks. In addition to cultural barriers towards the low-maintenance "urban forest" aesthetic championed by the park, local advocates have raised concerns regarding the displacement of communities in the area.

Project documentation claims that the current redevelopment of the area has increased Bangkok's green public spaces available to the community, strengthened the city's capacity to address climate hazards, improved air and soil quality, as well as provided water management benefits (UNA 2023). The degraded conditions of the area before the project posed a threat to the restoration of the landscape and there are ongoing efforts to set monitoring systems in place to evaluate the ecological benefits achieved by the park.

Writing: Erich Wolff and Perrine Hamel
Review: Carole Alazki
Copy Editor: Veronica Olivotto
Design: Phuong Anh Bui

Benchakitti forest park's wetlands and elevated walkways in Bangkok, Thailand.

References:

1. Huu Loc, Ho, Kim N. Irvine, Asan Suwanarit, Pakorn Vallikul, Fa Likitswat, Alisa Sahavacharin, Chansopheaktra Sovann, and Le Song Ha. 2020. "Mainstreaming Ecosystem Services as Public Policy in South East Asia, from Theory to Practice." In Sustainability and Law, edited by Volker Mauerhofer, Daniela Rupo, and Lara Tarquinio, 631–65. Cham: Springer International Publishing. Available at: https://doi.org/10.1007/978-3-030-42630-9_31 (Last Accessed: June 2025)

2. Linh, N.S., F. Ahmed, and H. Hu Loc. 2022. "Applications of Nature-Based Solutions in Urban Water Management in Singapore,Thailand and Vietnam : A Review." In Regional Perspectives of Nature-Based Solutions for Water: Benefits and Challenges, edited by Nevelina Pachova, Perlie Velasco, Antonina Torrens, and Veeriah Jegatheesan. Applied Environmental Science and Engineering for a Sustainable Future. Cham: Springer International Publishing. Available at: https://doi.org/10.1007/978-3-031-18412-3 (Last Accessed: June 2025)

3. UNA. 2023. "Benchakitti Forest Park." Urban Nature Atlas. Benchakitti Forest Park. May 2023. Available at: https://una.city/nbs/bangkok/benchakitti-forest-park (Last Accessed: June 2025)

4. Likitswat, Fa, and Chatchanin Sung. 2025. "Implementing Nature-Based Solutions: Learning from Benchakitti Forest Park in Bangkok." Journal of Architectural/Planning Research and Studies (JARS) 22 (2): 272299. Available at: https://doi.org/10.56261/jars.v22.272299 (Last Accessed: June 2025)

Photo by Erich Wolff.

AS_02 River Renaturalization in Bishan Ang Mo Kio Park

Context Overview

LOCATION

NBS TYPOLOGY	ECOSYSTEM TYPE
Restoration; Constructed & novel ecosystems	Wetlands; Rivers & waterways; Green spaces (gen.)
IMPACT SCALE	**URBAN TYPOLOGY**
Neighbourhood/Community	Urban core
SOCIAL CHALLENGES ADDRESSED	**CO-BENEFITS**
Physical and mental health; Flooding	Recreational benefits; Water quality; Reinforcement of socio-ecological knowledge; Increased biodiversity and ecosystem functions;

Bishan-Ang Mo Kio Park is a prominent urban green park in Singapore that has served as a regional example and catalyst for river renaturalization projects in Asia and globally. Situated at the southernmost tip of the Malaysian peninsula, Singapore is an island city-state and one of the most densely populated countries globally.

Bishan-Ang Mo Kio Park spans 62 hectares along Ang Mo Kio Avenue 1 and acts as a buffer around the Kallang River. Recognized as one of central Singapore's largest urban parks, it combines water features and lush greenery. As part of PUB's Active, Beautiful, and Clean (ABC) Waters programme, the former concrete canal was transformed into a naturalized 3 km meandering river, lined with flourishing wildflowers.

Introduction to the case study

Initially engineered in the 1960s, the area was characterized by a concrete-lined canal that contained the Kallang River and managed stormwater from the rapidly urbanizing Bishan and Ang Mo Kio satellite towns. In 1988, the park was developed around the canal, but integration was limited due to safety concerns and high-water velocities during storms.

A transformative shift occurred in the park's design philosophy. Rather than upgrading and restricting access to the concrete channel, the new approach utilized the parkland as a floodplain, as suggested by Rinaldi (2013). This innovative design not only enhances recreational opportunities with lush riverbanks during low water periods but also serves as flood buffer zones during high water levels, mitigating safety concerns.

Nature-based Solutions characteristics

Soil bioengineering integrates civil engineering design principles with natural materials like rocks and plants, as well as relying on plant roots for soil stabilization (Kallang River Bishan Park, 2014). To validate the effectiveness of these techniques in slowing water flow and preventing erosion, a 60-meter stretch of the river was redirected to serve as a test bed (PUB, Singapore, 2013c). While these methods had been applied globally, this marked one of the first implementation attempts in a tropical setting like Singapore. The test area aimed to demonstrate that riverbanks could be stabilized without resorting to conventional concrete walls (Rinaldi, 2013).

The park utilizes a cleansing biotope, an artificial wetland with 15 cells and four terraces designed for purifying river water and runoff (PUB, Singapore, 2014b). Nutrient-poor substrates in these cells host specific plants like Equisetum hyemale, Cyperus alternifolius, Portulaca oleracea, and Cyperus haspan, effectively cleansing water of contaminants and filtering particulates. Following passage through multiple wetland cells, the water undergoes ultraviolet treatment to eliminate bacteria before being reintroduced to the pond system, the park's river, or directed to a water playground within the park (NParks, Singapore, 2013).

Governance

The project aligns with Singapore's government vision for a "City in Nature" and with the previous policy that framed Singapore as the "Garden City". Singapore's national water agency (PUB) and the National Parks Board (NParks) jointly spearheaded the project, which served as a flagship initiative within PUB's ABC Waters program. The project, spanning from 2009 to 2012, incurred a total cost exceeding Euro 4,000,000 (USD 5,560,000) and was funded through the public national budget. Since its inauguration, the project has served as a flagship initiative to implement and disseminate river renaturalization and flood risk reduction strategies by the Singaporean government.

Outcome

The restored design offers a multifunctional space, allowing park goers to engage with the water and significantly improving flood prevention. The current channel can accommodate 40% more water than the previous channel (Kallang River Bishan Park, 2014). The river's meandering course extended the park's channel from 2.7 km to 3.2 km, contributing to slowed water flow, improved water treatment, filtration, and overall safety (Rinaldi, 2013). While the park effectively contributes to flood risk reduction by slowing down rainwater, risk communication is a critical component to ensure the safety of the community. The park uses a series of sensors, sound alarm systems and clear signage to communicate the risk that rising water levels could pose to park goers during rainy days.

Bishan-Ang Mo Kio Park showcases the successful use of bioremediation and soil bioengineering for stormwater and flood management in an urban tropical setting. The park employs vegetated bioswales, natural drainage channels on gentle slopes with rocks and plants to slow water flow and capture sediment. While not fully meeting stormwater treatment or water quality goals for nutrients, they excel at extracting coarse sediments and crucially provide pre-treatment for downstream systems like wetlands and bioretention systems.

Bishan Ang Mo Kio park.

Typical concrete canal used for urban drainage in Singapore similar to the one that existed in Bishan Ang Mo Kio park (left), renaturalized riverbed at Bishan Ang Mo Kio park after 2013 (right)

Writing: Ho Huu Loc and Erich Wolff
Review: Carole Alazki
Copy Editor: Veronica Olivotto
Design: Phuong Anh Bui

References:

1. Burris, L. (2006) People of the Poudre : an Kallang River Bishan Park. (2014). World Buildings Directory. Retrieved from http://www.worldbuildingsdirectory.com/project.cfm?id=3870.

2. NParks, Singapore. (2013). A Guide to Pond Gardens at Bishan-Ang Mo Kio Park. National Parks Board, Singapore. Retrieved from: http://www.nparks.gov.sg/cms/docs/PondGardens_070612.jpg.

3. PUB, Singapore. (2013b). Our Water, Our Future. Public Utilities Board, Singapore. Retrieved from: http://www.pub.gov.sg/mpublications/OurWaterOurFuture/Pages/default.aspx.

4. PUB, Singapore. (2013c) Kallang River @ Bishan-Ang Mo Kio Park Test Bed [Government]. Retrieved from: http://www.pub.gov.sg/abcwaters/ExploreABCAroundYou/Pages/KallangRiver-BishanPark2.aspx.

5. Rinaldi, B. M. (2013). The soft sell: in Singapore, Atelier Dreiseitl has given back a river people can love [BishanAng Mo Kio Park]. Landscape Architecture Magazine, 103(1), 76–89.

Photos by Erich Wolff.

AS_03 Chongqing Urban Regeneration Program

Context Overview

LOCATION

NBS TYPOLOGY	ECOSYSTEM TYPE
Restoration; Constructed & novel ecosystems	Green spaces (gen.); Wetlands; Stormwater infrastructure
IMPACT SCALE	**URBAN TYPOLOGY**
City	Urban core
SOCIAL CHALLENGES ADDRESSED	**CO-BENEFITS**
Flooding; Nature connection	Recreational benefits; Place regeneration & urban revitalization; Increased biodiversity and their functions; Water quality

Located in southwest China on the upper reaches of the Yangtze River, Chongqing is surrounded by several green mountain ranges, including the Daba, Wu, and Wuling mountains. It is also crossed by the Yangtze River, which is joined by the Wu River, the Min River, and major tributaries, stretching 691 km across the region and joining the Jialing River in its central area. The original site of the Dafosi Wetland Park in Tongnan consisted mainly of tidal flats carved out by the Fu River, primarily consisting of sand and gravel, but the river was later turned into a concrete-bottom canal.

Introduction to the case study

The case study of the Chongqing urban regeneration program exemplifies how a megacity built in a mountainous terrain can use nature-based solutions to provide benefits to its inhabitants (Xingzhong 2018). The central urban area of Chongqing is located in a typical ridge and valley region with high ecological sensitivity and intensive human activities. The pressure on the environment and natural resources is enormous, as 25% of the city's population lives in 5% of its land area and produces 43% of its GDP. In the process of rapid urbanization, conflicts between humans and nature are increasing, and the downtown or central areas are becoming overcrowded. In addition, urban development is conventionally dominated by the use of concrete buildings with impervious surfaces, increasing the risk of disasters such as floods and landslides.

Nature-based Solutions characteristics

Through the adoption of a "multi-landform" design strategy and innovative architecture, the project has applied nature-inspired design methods honoring the characteristics of a mountain city. The project adopted green infrastructure techniques such as construction of infrastructure in multiple levels, to generate a landscape capable of integrating human and ecological functions by reconciling transport needs and ecological corridors (Xingzhong 2018). Unused spaces around infrastructure corridors have been transformed into green corridors allowing the city to increase its canopy cover within its limited area. The Dafosi Wetland park was also renaturalized by removing concrete sections of the canal and implementing water retention ponds. Through the use of this approach, the park restored the habitats of the mudflats for flora and fauna, built ecological river embankments, and created riverside wetland parks in urban areas. The ongoing efforts to regenerate Chongqing build upon earlier greening initiatives, which can be traced back to the implementation of Chongqing Central park in 2013 and the regeneration of the Jiulongpo District with improvements to nine parks to operate as "sinks that are positive for both mitigation and adaptation" (World Bank 2018, 8).

Wetland vegetation used in water retention ponds in China.

Governance

The projects that comprise Chongqing's regeneration plan were primarily funded and maintained by the Chongqing Municipal People's Government with support from external funders including the World Bank, which financed 264 million US dollars for the Chongqing New Urbanization Pilot and Demonstration Project (World Bank 2018). The management of the parks is part of the city's portfolio of investments in greening initiatives. The governance structure of the project considered that the districts are responsible for developing their own community-led activities for environmental and economic regeneration (World Bank 2018, 5).

Outcome

In total, 22 national wetland parks have been established in Chongqing, and the protected wetlands cover an area of 207,200 ha, recording 563 species of vertebrates and 707 species of higher plants. As part of efforts to restore urban ecology, a total area 15,834.55 ha was reserved for urban parklands, 45% of which are located in central districts. 92 community sports and cultural parks were built on the undeveloped urban areas to meet the leisure and exercise needs of nearly 3.5 million residents. 1,198.9 ha of green spaces on slopes and in the central districts were reserved for urban greening. As of January 2021, the city has traded a total of 12,800 ha of forest cover quotas through a "green credits" mechanism, with a total trading volume of CNY 480 million (USD 72 million). By the end of 2019,

Chongqing has established the horizontal ecological compensation mechanism for 19 secondary rivers with a watershed area of ≥ 500 km2 and covering ≥ 2 districts and counties, with a total compensation fund of CNY 96.918 million.

Writing: Baolong Han
Review: Carole Alazki
Copy Editor: Veronica Olivotto
Design: Phuong Anh Bui

References:

1. Xingzhong, Y., Jia, Y., Lei, G. et al. (2018). Demonstration study on ecological restoration and landscape optimization of urban riverside littoral zone in the Three Gorges Reservoir area. Shanghai Urban Planning Review, (6):132-136. [InChinese] https://d.wanfangdata.com.cn/periodical/shcsgh201806035

2. World Bank (2018). "Chongqing New Urbanization Pilot and Demonstration Project (P158142)." Combined Project Information Documents/ Integrated Safeguards Datasheet (PID/ISDS) PIDISDSA23821. https://documents1.worldbank.org/curated/en/281131530860781079/pdf/Project-Information-Document-Integrated-Safeguards-Data-Sheet-Chongqing-New-Urbanization-Pilot-and-Demonstration-Project-P158142.pdf

Photo by Erich Wolff.

AS_04

The BORASSUS Project: Palm Geotextile Mats for Erosion Prevention

Context Overview

NBS TYPOLOGY Constructed & novel ecosystems	ECOSYSTEM TYPE Rivers & waterways; Riparian areas
IMPACT SCALE Greater urban region	URBAN TYPOLOGY Outside the city with urban impacts
SOCIAL CHALLENGES ADDRESSED Land degradation; Soil quality	CO-BENEFITS Green jobs and economic benefits; Water quality

Geotextile mats constructed from palm leaves are a historic example of how nature-based solutions have been used to stabilize soils and protect freshwater ecosystems in peri-urban agricultural areas. Key areas that have implemented this solution include the Huong Non commune (near Hanoi) in Vietnam and Ladkrabang (near Bangkok) in Thailand. These areas are located in watersheds that supply water to some of the biggest cities of Vietnam and Thailand (Fullen, 2009). The Red and Mekong watersheds, where the geotextiles were implemented within the BORASSUS project, house over 60 million people and experience a tropical savanna climate influenced by the Asian monsoon system, with a rainy season until October and a dry northeast monsoon until February.

Introduction to the case study

Rice farming is vital in Vietnam and implementing sustainable land management practices is crucial for maintaining river health, fertile soil and protecting riverine and coastal ecosystems with significant implications for cities. Recent advancements in soil conservation technology have significantly improved our understanding of soil erosion mechanisms over the last three decades. The use of biodegradable geotextiles in Vietnam under the BORASSUS project demonstrates the potential of applying palm fiber woven mats to address socio-economic issues through erosion control, poverty alleviation, and peri-urban employment. Despite synthetic geotextiles dominating the market in the region, organic geotextiles have been regarded as an effective erosion control strategy for decades, as highlighted by Langford and Coleman (1996) and Ogbobe et al. (1998).

Nature-based Solutions characteristics

Geotextiles made from *Borassus aethiopum* and *Mauritia flexuosa* leaves, termed Borassus and Buriti mats, respectively, offer cost-effective erosion control solutions. The BORASSUS project's experiences in Vietnam demonstrated their applicability to peri-urban agriculture, emphasizing their availability, simplicity, and immediate effectiveness at preventing the siltation of rivers and slowing down the loss of soil (Davies et al., 2006). The use of palm-mat geotextiles in Vietnam have been credited for serving as an income source to disadvantaged groups, supporting small and medium enterprises, promoting environmental education, and engaging local communities in environmental restoration (Fullen et al., 2007).

Governance

The BORASSUS Project was coordinated by the University of Wolverhampton and received EUR 1,749,876 (USD 2,082,352) funded by the European Commission (Contract INCO-CT-2005-510745). Running from 2005 to 2008, this research project implemented and evaluated the environmental and socio-economic impact of palm geotextiles for soil conservation and income generation. The project included a team spanning academic, government and community-led institutions in Europe, Africa, South-East Asia, and South America and was responsible for implementing the technology in different sites globally, including case studies in Vietnam and Thailand. While palm leaves and other organic materials have been used for construction for millenia, the project tested and documented for the first time its effects for controlling soil erosion. The implementation of geotextiles in sloping terrain in Vietnam involved local authority officers and farmers in the Tam Nong District and remunerated local residents for their efforts constructing mats (Fullen, 2009).

Outcome

The implementation of the project in Vietnam demonstrated that the use of biodegradable woven mats is effective at conserving soil moisture during the dry season and controlling erosion (preventing the loss of nutrients and soil) during the floods period (Fullen, 2009). The sustainable use of woven mats has encountered cultural barriers as some communities were not used to this practice or to the use of certain species of palms before the project. Despite these challenges, the project supported reforestation and agroforestry in peri-urban areas, providing financially deprived farmers with supplementary income through palm geotextiles. The development of palm geotextiles can create a labor-intensive industry in historically marginalized areas, improving the water quality in urban centers located downstream and slowing down the migration to cities by creating jobs in communities living in the urban periphery.

Writing: Ho Huu Loc and Erich Wolff
Review: Carole Alazki, Veronica Olivotto
Copy Editor: Veronica Olivotto
Design: Phuong Anh Bui

References:

1. Bhattacharyya, R., Fullen, M.A., Davies, K., Booth, C.A., 2009. Utilizing palm leaf geotextile mats to conserve a loamy sand soil in the United Kingdom. Agriculture Ecosystems & Environment 130, 50–58.

2. Davies, K., Fullen, M.A., Booth, C.A., 2006. A pilot project on the potential contribution of palm-mat geotextiles to soil conservation. Earth Surface Processes & Landforms 31, 561–569.

3. Fullen, M. (2009). The Environmental and Socio-economic Contribution of Palm Geotextiles to Sustainable Development and Soil Conservation (BORASSUS) Final Report (Report from 01/07/2005 to 28/02/2009). School of Applied Sciences, The University of Wolverhampton. https://cordis.europa.eu/project/id/510745/reporting/it

4. Fullen, M.A., Booth, C.A., Sarsby, R.W., Davies, K., Kugan, R., Bhattacharyya, R., Subedi, M., Poesen, J., Smets, T., Kertesz, A., Toth, A., Szalai, Z., Jakab, G., Kozma, K., Jankauskas, B., Jankauskiene, G., Bühmann, C., Paterson, G., Mulibana, E., Nell, J.P., van der Merwe, G.M.E., Guerra, A.J.T., Mendonça, J.K.S., Guerra, T.T., Sathler, R., Bezerra, J.F.R., Peres, S.M., Yi, Z., Yongmei, Li, T., Panomtarachichigul, M., Peukrai, S., Thu, D.C., Cuong, T.H., Toan, T.T., Jonsyn-Ellis, F., Jallow, S., Cole, A., Mulholland, B., Dearlove, M., Corkill, C., (2007). Contributions of biogeotextiles to sustainable development and soil conservation in developing countries: the BORASSUS Project. Ecosystems and Sustainable Development. WIT Press, Southampton, UK, pp.123–141.

5. Langford, R.L., Coleman, M.J., (1996). Biodegradable erosion control blankets prove effective on Iowa wildlife refuge. XXVII Int. Erosion Control Assoc. Proceedings of Conf., Seattle, USA, pp. 13–20.

6. Ogbobe, O., Essien, K.S., Adebayo, A., (1998). A study of biodegradable geotextiles used for erosion control. Geosynthetics International 5, 545–553.

AS_05 Pingshan River Water Treatment Wetlands

Context Overview

NBS TYPOLOGY	ECOSYSTEM TYPE
Constructed & novel ecosystems	Wetlands; Rivers & waterways
IMPACT SCALE	**URBAN TYPOLOGY**
Greater urban region	Metropolitan
SOCIAL CHALLENGES ADDRESSED	**CO-BENEFITS**
Water quality	Increased biodiversity and their functions; Flood mitigation; Recreational benefits; Green jobs and economic benefits

The Pingshan District lies in the northeast of Shenzhen, Guangdong Province and has a subtropical-oceanic climate with an average annual temperature between 20 and 28 °C and 1,705 mm of rainfall. The Pingshan watershed underwent a significant transformation after the implementation of water treatment wetlands in 2011. Located in the northeast of Shenzhen City, in the Guangdong Province of China, the Pingshan watershed has a subtropical oceanic climate and a population of 428,000 people. The Pingshan River Basin is highly urbanized and is characterized by low water levels with high amounts of sediment that were affected by wastewater and industrial pollution.

Introduction to the case study

The Pingshan River Basin was previously dominated by industrial activity, whose effluents were released directly into the river making it heavily polluted. From 2011 onwards, the industrial complexes started to move to other locations creating an opportunity for the restoration of the river. In the following years, eight vertical-flow treatment wetlands were gradually installed to improve water quality through water treatment and restore the ecosystem of the Pingshan River Basin (Zhai et al. 2021). The resulting product from the wetlands is safe to be released in the river ecosystem and serves as another source of water for other activities.

Nature-based Solutions characteristics

The solution adopted is a low cost nature-based solution with co-benefits for the residents, as the riverbanks are now a recreational area. The wetlands have increased the biodiversity in the Pingshan River Basin by removing pollutants and enhancing ecological function. The design of the wetlands is also aligned with the plans set by the municipality to reach 30% of vegetated cover in the area. The wetlands have varying sizes and dimensions ranging from 1.76 hectares to 12.8 hectares, while the total area covered by the whole complex is 50 hectares. The wetlands use a combination of plants with bioremediation potential, including Cyperus alternifolius, Pontederia cordata and Cyperus papyrus. The treated water from the wetlands is redistributed to different parts of the river, where it enters "an ecological purification zone" with aquatic ecosystems that are accessible through an ecological park (Zhai et al. 2021). The ecological purification zone is described as a series of retention basins, ponds and streams with vegetation adapted to both tidal and submerged conditions, further improving the retention of sediments and enhancement of water quality.

Governance

The project was funded using resources from the People's Government of Pingshan District since 2018, which has invested more than 7 billion yuan to complete the regulation of the tributaries of Pingshan River. Contributing to national goals and aligned with the vision of the wider Sponge Cities program, this is an ongoing project under the management of the Pingshan District Water Affairs Bureau, which oversaw the construction and is in charge of the wetland planning.

Outcome

The eight treatment wetlands along the Pingshan River Basin help to reduce pollutant concentrations in the water before it enters the river, thus improving the river's water quality. Climatic uncertainties were considered a significant challenge for implementing the wetlands as the system had to continue operating to control water pollution regardless of varying water flows and seasonality. Despite the varied temperatures and rainfall throughout the year, evaluation has demonstrated that the wetlands meet the quality standards according to the area's "Environmental quality standards for surface water" (Zhai et al. 2021). The wetlands have fulfilled the environmental goals set by the project and led to benefits for wildlife, including an increased number of pollinators and higher biodiversity in freshwater ecosystems. Among other co-benefits, the wetlands are expected to serve as areas of cultural value and social bonding, while also providing flood regulation, stormwater control and carbon sequestration services (Zhai et al. 2021).

Wetland systems can significantly improve water quality and enhance the recreational potential of rivers in China.

Writing: Baolong Han
Review: Carole Alazki
Copy Editor: Veronica Olivotto
Design: Phuong Anh Bui

References:

Zhai, Jun, Wenbo Liu, Gu Huang, and Lobna Amin. 2021. "Vertical Flow Treatment Wetlands for Pollution Control in Pingshan River Watershed, Shenzhen, China." In Nature-Based Solutions for Wastewater Treatment: A Series of Factsheets and Case Studies, edited by Katharine Cross, Katharina Tondera, Anacleto Rizzo, Lisa Andrews, Bernhard Pucher, Darja Istenič, Nathan Karres, and Robert McDonald, 1st ed. London, United Kingdom: IWA Publishing. Available at: https://doi.org/10.2166/9781789062267. (Last Accessed: June 2025)

Photo by Erich Wolff.

AS_06 Shenzhen Mangrove Restoration

Context Overview

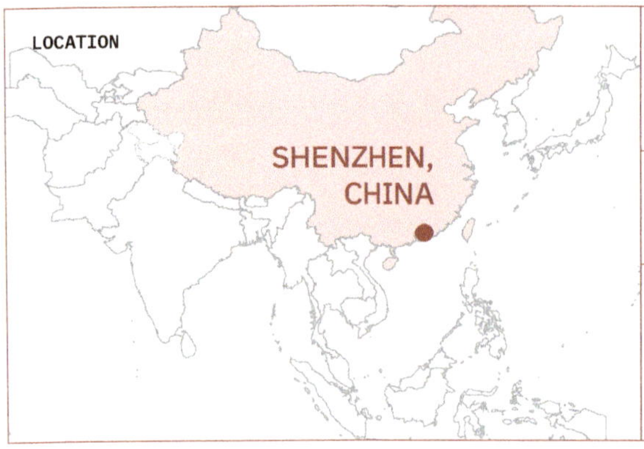

NBS TYPOLOGY Protection & conservation; Restoration	ECOSYSTEM TYPE Mangroves; Lakes & ponds; Coastal (gen.)
IMPACT SCALE Greater urban region	URBAN TYPOLOGY Peri-urban
SOCIAL CHALLENGES ADDRESSED Land degradation; Nature connection; Biodiversity loss	CO-BENEFITS Urban cooling; Carbon sequestration; Water quality

Bordering both Shenzhen and Hong Kong, two of China's largest metropolitan areas, the wetlands of Shenzhen Bay are an important example of the potential of nature-based solutions for the conservation and restoration of ecosystems. They are wintering habitat and stopover for one of the world's nine major bird migratory routes - the East Asian-Australasian Flyway - with some 100,000 migratory birds wintering and passing through each year (Lili, et al. 2019). To effectively conserve this mangrove wetland ecosystem close to the center of the megacity, the Shenzhen government has launched a series of coastal mangrove restoration policies in Shenzhen Bay.

Introduction to the case study

The rapid increase in urbanization in Shenzhen has been accompanied by a decrease in the mangrove areas. Direct industrial and municipal wastewater discharges have increased harmful pollutants and wetland degradation. Along the coastal estuary, watercourses have been converted into channels, preventing the exchange of matter and energy between terrestrial and aquatic ecosystems. Diked fishponds have also been degraded, reducing the ecological function of providing habitats and feeding grounds for migratory birds. Invasive plant species such as Mikania micrantha and Leucaena leucocephala have taken over large areas, while outbreaks of insect infestations have become more frequent, and local species have been replaced across ecological niches, all of which have led to an oversimplified and increasingly fragile structure of wetland communities.

Nature-based Solutions characteristics

The restoration and conservation of mangrove wetlands in Shenzhen is being implemented to improve the disaster risk reduction and climate change adaptation capacity of this megacity (Zhigang, et al 2019). The first step was to renaturalize the watercourses, soften the course of the Fengtang River, and reconstruct the banks of the river and its branches, removing rigid structures from the surface, lowering the water level, and reforming the side slopes into a gentle gradient. Organic matter was reintroduced to the side slopes to restore the soil environment for better plant growth. Secondly, according to the ecological characteristics of the environment from the outer tidal flats to the inner banks, various plant communities were restored to rehabilitate the bank vegetation, and a comprehensive ecosystem of mangroves, associate mangroves and riverine plants was developed, contributing to the restoration of both the main mangrove wetland functions and the estuarine ecosystem of the Fengtang River.

Governance

Learning from the experience of Mai Po Nature Reserve in Hong Kong (Jun, Y. 2019), the Futian Reserve Management Office manages the project in cooperation with the Shenzhen Mangrove Wetlands Conservation Foundation (MCF, IUCN member). Futian Mangrove Ecological Park is an ecological park funded by the Shenzhen Municipal People's Government and operated by the non-governmental organization MCF (Lina, et al. 2013).

Outcome

The project includes mangrove conservation and sustainable management, which have led to an increase in the total area of mangrove forests and reversed the trend of mangrove wetland ecosystem degradation. Within the restoration area, mangrove wetlands have recovered and biodiversity has shown a consistent increase with a remarkable growth in both the number and diversity of species, particularly following the restoration of fish ponds into bird habitats. Through a series of successive restoration actions, the project is playing an important role in maintaining the ecological status and value of Shenzhen Bay on migratory bird flyways.

Results have shown that in Shenzhen Bay, each hectare of mangrove forests absorbs about 40 t CO_2 per year from the atmosphere, highlighting the crucial role mangroves play in carbon fixation, mitigating climate change, and helping achieve carbon neutrality (Congjiao, et al. 2016). Moreover, the wide and open water surface of Shenzhen Bay has had vast areas of mangrove forests restored. Accordingly, the cooling effect and direct regulation of climate contributed by water vaporization and plant transpiration greatly mitigate the urban heat from nearby Shenzhen.

The mangrove forest ecosystem restoration project has succeeded in maintaining the harmony and connectivity between the restoration zone and the surrounding areas, while establishing a natural link between the sea and the city, and between birds and people, thus promoting the functions of coastal wetland ecosystems. At present, the coastal mangrove wetlands in Shenzhen Bay serve as a demonstration of how nature-based solutions can protect a famous scenic spot for citizens and visitors at home and abroad, providing services such as sightseeing, leisure and science dissemination to more than 10 million visitors every year.

Writing: Baolong Han
Review: Carole Alazki
Copy Editor: Veronica Olivotto
Design: Phuong Anh Bui

References:

1. Jun, Y. (2019). Shenzhen Bay becomes clear, how to implement in Futian? China Ecological Civilization, (4): 61-63. [In Chinese]
2. Zhigang, Q., Xi, Y., Lingyun, Y. and et al. (2019). Ecological problems and conservation strategies for mangroves under impacts of urbanization: A case study of futian, shenzhen. Wetland Science and Management, 15 (3): 31-34. [In Chinese]
3. Lili, S., Yunzhu, L., Yifei, J. et al. (2019). Comparison of diversity of waterbird communities before and after ecological restoration of fishing ponds in Guangdong Neilingding Island-Futian National Nature Reserve. Wetland Science, 17 (6): 631-636. [In Chinese]
4. Lina, L., Lie, C., Pingping, W. et al. (2013). A study on the ecological problems and restoration measures of Futian Mangrove Nature Reserve in Shenzhen. Ocean Technology, 32(2): 125-132. [In Chinese]
5. Congjiao, P., Jiawei, Q., Xudong, G., et al. (2016). Vegetation carbon stocks and net primary productivity of the mangrove forests in Shenzhen, China. Chinese Journal of Applied Ecology, 27(7): 2059-2065. [In Chinese]

AS_07 Urchinomics: Urchin Harvesting for Marine Restoration

Context Overview

NBS TYPOLOGY	ECOSYSTEM TYPE
Restoration	Productive blue spaces
IMPACT SCALE	**URBAN TYPOLOGY**
Transboundary (subnational)	Metropolitan
SOCIAL CHALLENGES ADDRESSED	**CO-BENEFITS**
Biodiversity loss	Carbon sequestration; Green jobs and economic benefits

Responding to the pressing issue of urchin overpopulation in coastal urban areas, the technology used in Urchinomics illustrates the potential of removing invasive species as a nature-based solution to restore overgrazed marine ecosystems in coastal areas of Asia and beyond (Eger et al., 2022). By controlling the overgrazing of marine urban ecosystems, the project has successfully restored large areas of kelp forests near the company's facilities in Kunisaki and Nagato, in Japan (Ocean Decade Network, 2022b). The project is currently expanding its operations beyond Japan and now has active facilities in other continents. Feasibility studies are also being conducted in several other contexts in Asia, including in the coasts of Korea and China.

Introduction to the case study

Urchinomics is a private company based in Japan championing the restoration of marine environments through the removal and commercialization of urchins. The project promotes the management of marine ecosystems by employing commercial divers to harvest urchins from overpopulated areas. The removal of urchins plays an important role in ecological restoration as it kickstarts the reestablishment of kelp forests otherwise unlikely in overgrazed marine areas (Eger et al., 2022). The removal of urchins leads to the production of marine biomass, increased biodiversity, and carbon sequestration (Ocean Decade Network, 2022a).

Nature-based Solutions characteristics

The creators of Urchinomics describe the initiative as an aquaculture venture that aims to restore marine ecosystems by controlling the impacts caused by urchins through the harvesting of grazer populations (Ocean Decade Network, 2022a). The commercialization of sea urchin roe entails the manual collection of urchins from degraded ecosystems by divers and the ranching of the animals in land-based aquaculture facilities (Eger et al., 2020). This step is necessary to make the commercialization of urchins economically viable due to the poor quality of the roe found in urchin barrens, where overgrazing makes the environment devoid of kelp forests and other food sources (Eger et al., 2020).

Some authors argue that making this operation economically viable is critical to support the ecological restoration of urchin barrens, as these ecosystems are unable to revert back into previous conditions without external intervention. In these conditions, barren ecosystems can persist indefinitely as urchins can survive under starvation for decades (Eger et al., 2020).

Governance

The effects of the project (including its urchin harvesting, and kelp forest restoration efforts) can be monitored using indicators, which include keeping track of urchin and kelp densities in the short term; measuring the population expansion and evaluating the demographic structure of the forest in the medium term; and assessing these parameters as well as biomass in the long term. The project has recently been listed as a certified "voluntary carbon credit deal" by the Japan Blue Economy Association (JBE), indicating that active monitoring efforts are being undertaken both by Urchinomics and by external observers (Ocean Decade Network, 2022b). The financing and governance of the project is currently tied to its promises in terms of achieving blue carbon offsetting goals. Japan's largest company in the field of energy, ENEOS, has financially supported Urchinomics in order to fulfill its plans to decarbonize its operations by 2040. Beyond demonstrating interest in purchasing carbon credits from Urchinomics, ENEOS has sponsored research and development of the company's business model as a third-party certified blue carbon credit provider. The profits from the operation of this business model also mean that this project has additional social and economic benefits, such as the generation of new income sources and creation of employment opportunities. The recovery of kelp forests has, therefore, ecological benefits while also addressing societal challenges.

Outcome

The method employed by Urchinomics has been endorsed by the United Nations committee for the Decade of Ocean Science for Sustainable Development (2021-2030) (Ocean Decade) as one of three commercial initiatives contributing to the restoration of marine ecosystems (Ocean Decade Network, 2022a). The project operates as a private business considering that the urchin removal strategy is made viable by the commercialization of these animals as a luxury food item, generating sustainable revenues (Eger et al., 2022). The results of these types of projects in Japan are underrepresented in the global literature due to language barriers and commercial protections. While Japan has a long history of projects aiming to restore marine environments, many of the findings are kept as industrial intellectual property or are only published in Japanese.

The restoration of kelp forests in Asia improves biodiversity in coastal areas.

Writing: Erich Wolff and Perrine Hamel
Review: Carole Alazki
Copy Editor: Veronica Olivotto
Design: Phuong Anh Bui

References:

1. Eger, A. M., Marzinelli, E., Gribben, P., Johnson, C. R., Layton, C., Steinberg, P. D., Wood, G., Silliman, B. R., & Vergés, A. (2020). Playing to the Positives: Using Synergies to Enhance Kelp Forest Restoration. Frontiers in Marine Science, 7, 544. https://doi.org/10.3389/fmars.2020.00544

2. Eger, A. M., Marzinelli, E. M., Christie, H., Fagerli, C. W., Fujita, D., Gonzalez, A. P., Hong, S. W., Kim, J. H., Lee, L. C., McHugh, T. A., Nishihara, G. N., Tatsumi, M., Steinberg, P. D., & Vergés, A. (2022). Global kelp forest restoration: Past lessons, present status, and future directions. Biological Reviews, 97(4), 1449–1475. https://doi.org/10.1111/brv.12850

3. Ocean Decade Network. (2022a, July 1). Urchinomics: One of three commercial ventures in the world endorsed by the United Nations (UN) Ocean Decade. Ocean Decade. https://oceandecade.org/news/urchinomics-one-of-three-commercial-ventures-in-the-world-endorsed-by-the-united-nations-un-ocean-decade/

4. Ocean Decade Network. (2022b, November 22). Urchinomics secures world first kelp restoration blue carbon credits. Ocean Decade. https://oceandecade.org/news/urchinomics-secures-world-first-kelp-restoration-blue-carbon-credits/

Photo by Erich Wolff.

AS_08 Gangxia 1980 project: A Green Roof for Youths' Mental Health in Shenzhen

Context Overview

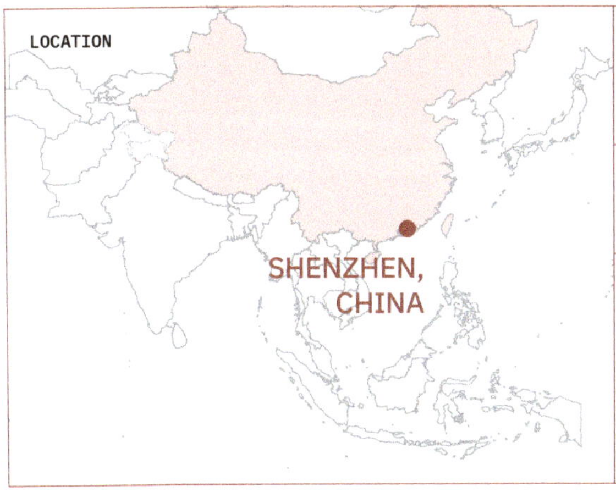

NBS TYPOLOGY Constructed & novel ecosystems	ECOSYSTEM TYPE Roofs & walls; Agriculture & food spaces
IMPACT SCALE Building	URBAN TYPOLOGY Urban core
SOCIAL CHALLENGES ADDRESSED Social wellbeing; Flooding	CO-BENEFITS Place regeneration & urban revitalization; Stronger human-nature connections; Recreational benefits

The Gangxia 1980 project is a cultural transformation project on the rooftops of old village buildings. Located in Shenzhen, one of the fastest urbanizing and newest cities in China, the Gangxia 1980 project illustrates the potential for nature-based solutions in dense urban areas. During the past 40 years, the population of Shenzhen has grown from 200,000 to 20 million, with the majority being young people. In 2020, the average age of residents in the city was 32.5 years, much lower than the national average of 38.8 years. Residents live in a built-up area of less than 1000 square kilometers. Most of Shenzhen's young people are in the early stages of their careers, with low incomes and living in low-rise villages under the shadow of skyscrapers (Wang, Y. 2019).

Introduction to the case study

As modern cities continue to expand, the natural villages originally located on the outskirts of the city have gradually become isolated islands in bustling metropolises. As a product of the rapid population growth in modern China, urban villages are widespread in major cities across the country. Despite being called villages, these settlements are quickly losing vegetation cover, which is being replaced by impervious concrete surfaces. This is also true for the urban villages in Shenzhen, where income disparities and busy work schedules place a significant psychological burden on young people. The rooftop greening project Gangxia 1980, aims to achieve two main objectives: to provide the residents of urban villages with green spaces for shared use and to improve water management at the local scale.

Nature-based Solutions characteristics

The 90 square meters of well-arranged rooftop greenery not only serves the function of collecting and reusing rainwater but also becomes a venue for cultural activities for nearby young people (UNA, 2023). Engineers have placed planting boxes on the rooftop, which conserve rainwater and allow for the cultivation of vegetables or ornamental plants. Over 400 planting boxes on the rooftop are used to collect rainwater, which is then connected to a 300-liter rainwater collection tank, allowing the rainwater to be stored in the boxes. At the same time, these planting boxes have been utilized to enhance the cultural aspects of this rooftop greening project.

Governance

The project was launched by The Nature Conservancy (TNC) in collaboration with Zhubo-AAO (a design institute), Glocal Estate Management (manages the project building and owns the renovation project), the Paradise International Foundation (a nonprofit environmental organization) and Urban Planning & Design Institute of Shenzhen. Due to the unique design of the Gangxia 1980 green roof, during the construction phase, some residents of Gangxia village thought illegal construction was going on and filed complaints to the district law enforcement, resulting in the project being forced to a halt. To resolve the problem, TNC made various visits to the property manager, to local community centers, government authorities and law enforcement. The team explained the project and its objectives in further detail, invited them to visit the site and shared details of the design plan with them. After several meetings, the project finally received approval from local authorities and the community (Qiang, V. & Yu, X. 2020). Currently, the management and maintenance are the responsibility of the property owner, and the maintenance costs are covered by the rent. Due to the popularity of the project, the rent for the property is slightly higher than the surrounding area.

Outcome

Through the planting and water storage equipment on the roof, a 65% annual runoff control rate for rainwater can be achieved. Since the garden's official opening, TNC has led two summer nature education programs for local children, where students from the village have had the opportunity to be outdoors and participate in educational activities. The program also uses the green roof for children to practice and observe nature in action as they try out their gardening skills. To help build relationships among neighbors, in June 2018 TNC also worked with the Sinolink Education Center in Shenzhen to host a live music concert on the green roof.

Writing: Baolong Han
Review: Carole Alazki
Copy Editor: Veronica Olivotto
Design: Phuong Anh Bui

References:

1. Qiang, V. & Yu, X. (2019). The Green Cloud Project: Finding Green Space in an Urban Village (Shenzhen, China). Biophilic Cities Journal. 32-35. Available at: Source link [Accessed: 2 October 2021].
2. Qiang, V. & Yu, X. (2020).The Green Cloud, A Rooftop Story from Shenzhen: A "Living" Sponge Space Inside an Urban Village. The Nature of Cities. Available at: Source link [Accessed: 2 October 2021].
3. Green Cloud/Zhubo-AAO. Arch Daily. Available at: Source link. com/902375/green-cloud-zhuboaao [Accessed: 2 October 2021].
4. Wang, Y. (2019). The practice of grassroots green roof project in Shenzhen, China. Master thesis, Radbound University. Available at: Source link [Accessed: 2 October 2021].
5. UNA. 2023. "The Green Cloud Project - Gangxia 1980" Urban Nature Atlas. The Green Cloud Project - Gangxia 1980. October 2021. https://una.city/nbs/shenzhen/green-cloud-project-gangxia-1980

AS_09 Shijiu Yang Wetland Park for Water Treatment

Context Overview

LOCATION

NBS TYPOLOGY	ECOSYSTEM TYPE
Constructed & novel ecosystems	Wetlands; Corridors; Stormwater infrastructure
IMPACT SCALE	**URBAN TYPOLOGY**
Transboundary (subnational)	Peri-urban
SOCIAL CHALLENGES ADDRESSED	**CO-BENEFITS**
Water quality; Water security	Human health and wellbeing

The Shijiu Yang Wetland Park, in the eastern part of China, is a remarkable example of how nature-based solutions can be used to provide multiple societal benefits while supporting biodiversity. In order to quickly solve poor drinking water quality for citizens of Jiaxing City, the local government intends to use the Xintengtang and Beijiao River as alternative water sources. The use of water from these rivers, however, requires additional treatment and, for this reason, the wetland park is located adjacent to a water supply plant to directly contribute to improve water quality in the region.

Introduction to the case study

With the rapid development of the economy and industry, the Taihu Lake (including the Xintengtang and Beijiao Rivers) basin has been severely polluted, leading to the deterioration of the water quality in Jiaxing and its surrounding areas. Located downstream of the Taihu Lake basin, Jiaxing's water resources mainly rely on the ecosystem health of the river upstream, which is compromised by degrading water quality, the severe siltation in tributaries, pollution sources, eutrophication and low water levels. The water quality in most parts of the river has been ranked as Class IV and Class V, almost the worst possible category (Sohu News, 2022).

Nature-based Solutions characteristics

The Shijiu Yang consists of a large vegetated wetland area that expanded the earlier wetland system. The original area in the western part of Shijiu Yang was transformed into a subsurface wetland system by adding substrate material and redirecting the natural water flow to improve the treatment potential. The main interventions in the area included the addition of a raw water pretreatment system, the construction of ecological filter dams and overflow weirs, improvement of roads and access, as well as changes in the wetland flow for ecological restoration purposes. The Shijiu Yang Wetland Park in Jiaxing City now contains a series of water bodies following the creation of new lakes and water channels, which are utilized to perform ecological functions by trapping and treating micro-pollutants. It also employs multi-level ponds and swales for additional treatment, forming a comprehensive connected park with wetland landscapes as its main feature, effectively alleviating the pressure on the water treatment plant (Jiaxing Shijiu Yang Provincial Wetland Park 2014).

Governance

In early 2005, a research team from the Research Center for Eco-Environmental Sciences of the Chinese Academy of Sciences conducted field surveys, formulated and finalized the plan in collaboration with the Jiaxing Municipal Committee and Government. The construction was organized by the Jiaxing Municipal Government, and Jiaxing Water Conservancy Investment Co., Ltd. was involved in the operation and maintenance.

Outcome

When a pollution incident occurs in the water intake channel, rendering water intake impossible, the water stored in the wetland can meet the water demand for 3-4 days. As a result, the wetland serves as a safety reserve and buffer for urban water plants. Monitoring data shows that the raw water provided by the Shijiu Yang wetland to the Shijiu Yang water supply plant has maintained a water quality of Class II for three consecutive years. This means that the overall water quality of the river network in Jiaxing City has gradually improved. The effluent water quality from the Shijiu Yang wetland reached an annual average of Class III for the first time in 2016, and for the first time reached Class II in June 2019, and has since remained stable at Class II water standards (Sohu News, 2022), only one level below the best possible water quality standards.

This project serves as an example of how wetland systems can contribute to the improvement of water quality in China's small-scale polluted water sources. The innovative and sustainable practices pioneered by Jiaxing City have received attention from relevant national ministries and commissions. In December 2011, the project received the "China Habitat Environment Model Award". At the same time, the project was recognized as one of the "Top 100 Global Examples" for the 2012 "Dubai International Best Practices Award for Improving the Living Environment" by the United Nations Human Settlements Programme (Sohu News, 2022).

Writing: Baolong Han
Review: Carole Alzaki
Copy Editor: Veronica Olivotto
Design: Phuong Anh Bui

References:

1. Jiaxing Shijiu Yang Provincial Wetland Park [J] (2014). Zhejiang Forestry, (S1): 86-87.

2. Sohu News (2022) Shijiu Yang Water Source Wetland: Pioneering in China, 'Jiaxing Experience' Attracts Attention [N].

AS_10 The Floating Gardens of Inle Lake in Myanmar

Context Overview

NBS TYPOLOGY	ECOSYSTEM TYPE
Restoration	Lakes & ponds; Wetlands
IMPACT SCALE	**URBAN TYPOLOGY**
Greater urban region	Peri-urban
SOCIAL CHALLENGES ADDRESSED	**CO-BENEFITS**
Water quality; Food security; Economic vulnerability	Increased biodiversity and their functions; Reinforcement of socio-ecological knowledge

The gardens of Inle Lake are an indigenous nature-based solution consisting of floating organic material that recycles nutrients and allows farming to become less dependent on seasonality, while providing a variety of food sources for farming families. The Inle Lake is a long, flat-bottomed lake located in the Shan State of Myanmar at an altitude of 880 m above sea level and is around 1.5 million years old.

The lake, second largest in Myanmar, is a significant freshwater biodiversity hotspot with local, national, and global socio-ecological significance. The 70,000 people who live around and on the lake are largely from the Intha cultural group. The Intha are associated with a hydroponic practice called yechan that entails the creation and maintenance of floating gardens that enable high productivity in both natural and artificial environments. The floating gardens cover a total area of 3833 ha (Oo et al. 2021). The lake's length is approximately 18 km from north to south and 6.4 km from east to west. The shallow freshwater section of the lake is 38 km2 while the entire lake area including stilt houses, wetlands, and floating gardens is approximately 128 km2.

The lake is home to several threatened bird species and is also a significant nesting and breeding ground for fishes and amphibians. Agricultural activities involving the lake are influenced by a monsoonal, humid, and subtropical local climate with an annual average temperature ranging from 23°C to 35°C.

Introduction to the case study

The Inle Lake in Myanmar gained visibility at the end of political isolation in Myanmar in 2011, which paved the way for the country to participate in global agricultural trade and international tourism (Peters et al. 2022). These activities have caused an intensification in the consumption and production patterns around the lake. In turn this increased the demand for food and natural resources, especially tomatoes, the area's cash crop grown on floating gardens that use yechan. Furthermore, the lake's ecosystems are threatened by climate risks, including severe droughts, heavy rainfall and flooding, and high weather variability. Fish stock species have also decreased over the decades mainly because of water pollution from fertilizer and pesticides and waste dumping (Oo et al., 2021).

Nature-based Solutions characteristics

The floating gardens of Inle Lake have been documented as early as 1918 and were recognized as an economic sector in 1960 when improvements in infrastructure enabled the expansion of agricultural activity through the gardens (Peters et al., 2022). The floating gardens are formed using grasses, reeds, and other aquatic vegetation. The floating roots intermingle and form a dense mass for the build-up of soil while silt gathers through the underwater roots (Oo et al. 2021). It takes 30 years for the mass to reach a thickness appropriate for farming. The vegetation is then cut into wide stripes of about two meters wide, the stripes are attached to boats and moved closer to the farmer village where they are staked to the lake bed and separated by two-meter-wide channels. While the floating gardens are largely associated with the production of tomatoes, farmers also grow bottle gourd, cucumber, yard-long beans, and flowers. Seeds for the floating gardens are grown near farmer's houses using compost soil, in addition to mud from the lake (Oo et al. 2021).

Governance

Aside from the farmers and environmental NGOs, national level agencies also administer the floating islands, among them, the Department of Agricultural Land Management and Statistics, the Department of Agriculture, the Department of Forest, and the Department of Tourism. Funding to ensure sustainable agricultural practices, including reduced use of chemical fertilizer, has come from institutions such as the International Finance Corporation and the Government of Norway. A 2012 program between UNDP and the Government of Norway sought to reverse environmental degradation through engaging 9,500 households in organic farming, waste management, fishery control, and other ecologically sound farming practices (UNDP 2019).

Outcome

As intensified agriculture and tourism practices have significantly strained the lake's ecosystems, sustainable farming interventions have contributed to the preservation of the lake, increased the quality of produce, and halted battery-shock fishing (International Finance Corporation 2018; UNDP, 2019). Meanwhile the UNDP program with Norwegian Government support, included the construction of check dams which managed the flow of streams towards the lake, thus reducing siltation and recharging the groundwater and providing a reliable source of water for paddy farmers (UNDP, 2019).

Writing: Pamela Cajilig
Review: Carole Alazki
Copy Editor: Veronica Olivotto
Design: Phuong Anh Bui

References:

1. International Finance Corporation. 2018. 'IFC Promotes Knowledge Exchange on Good Agriculture Practices to Support Inle Lake Tomato Farmers'. International organisation. International Finance Corporation. 2018. https://ifcpressreleasesprod.aseprod.ifc.org/all/pages/PressDetail.aspx?ID=17559.

2. Karki, Seema, Aye Myat Thandar, Kabir Uddin, Sein Tun, Win Maung Aye, Kamal Aryal, Pratikshya Kandel, and Nakul Chettri. 2018. 'Impact of Land Use Land Cover Change on Ecosystem Services: A Comparative Analysis on Observed Data and People's Perception in Inle Lake, Myanmar'. Environmental Systems Research 7 (1): 25. https://doi.org/10.1186/s40068-018-0128-7.

3. Peters, Kristin, Paul D. Wagner, Ei Wai Phyo, Win Win Zin, Cho Cho Thin Kyi, and Nicola Fohrer. 2022. 'Spatial and Temporal Assessment of Human-Water Interactions at the Inle Lake, Myanmar: A Socio-Hydrological DPSIR Analysis'. Environmental Monitoring and Assessment 195 (1): 220. https://doi.org/10.1007/s10661-022-10730-4.

4. Oo, Moe Thae, Zin Wai Aung, and Clelia Puzzo. 2022. 'The Floating Garden Agricultural System of the Inle Lake (Myanmar) as an Example of Equilibrium between Food Production and Biodiversity Maintenance'. Biodiversity and Conservation 31 (10): 2435–52. https://doi.org/10.1007/s10531-021-02347-9.

5. UNDP. 2019. 'Protecting Livelihoods and Nature on Inle Lake'. International organisation. UNDP. 14 November 2019. https://www.undp.org/myanmar/stories/protecting-livelihoods-and-nature-inle-lake.

AS_11 Sustainable Coastal Protection through Biodiversity Conservation in Coastal Ecosystems Affected by Typhoons in the Philippines (ProCoast)

Context Overview

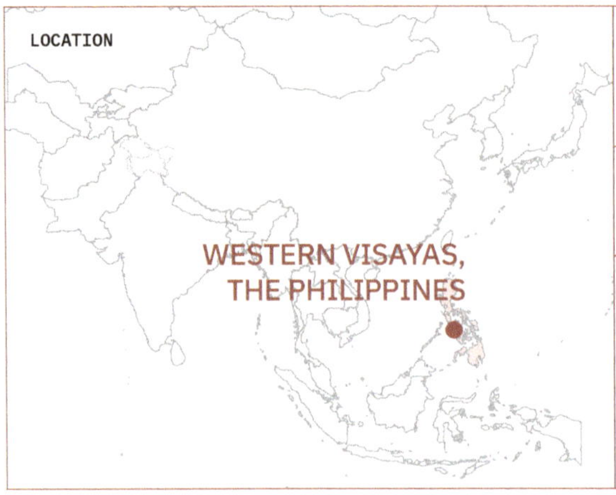

NBS TYPOLOGY Protection & conservation; Restoration	**ECOSYSTEM TYPE** Mangroves; Productive blue spaces
IMPACT SCALE City	**URBAN TYPOLOGY** Peri-urban
SOCIAL CHALLENGES ADDRESSED Sea level rise; Storms; Social inclusion	**CO-BENEFITS** Increased biodiversity and their functions; Reinforcement of socio-ecological knowledge

The ProCoast project is a remarkable example of how nature-based solutions can be used to reduce disaster risk in coastal cities. The project is timely because the Philippines has often been viewed as one of the countries most exposed to climate risks partly due to its location in monsoonal Southeast Asia. The country experiences as many as 20 typhoons annually and its largely coastal population also faces rising sea levels 5.7mm to 7.0mm annually, well above the global annual average rate of 3.2mm (Oscar M. Lopez Center, 2023).

The Philippines is regarded as one world's 20 countries with mega biodiversity and contains 46 of the world's 70 mangrove species. With 70% of the population living in coastal areas that are experiencing rapid and unsustainable urban and aquaculture development, the coastal ecosystems—mangroves, coral reefs, and seagrass beds—that once provided livelihoods and natural protection from coastal hazards, are now under pressure and in decline.

Introduction to the case study

The Sustainable Coastal Protection Through Biodiversity Conservation in Coastal Ecosystems Affected by Typhoons in the Philippines (ProCoast) project, aims to work at the national and local government level in reversing the depletion of mangrove and beach forests, coral reefs, and seagrass beds. By restoring these ecosystems, the project aims to help coastal communities by strengthening protection against coastal hazards such as typhoons and storm surges.

Mangrove restoration in Iloilo (the Philippines)

Mangrove area restored as part of PROCOAST activities in the Philippines.

Nature-based Solutions characteristics

Depending on the site, ProCoast uses a diverse set of methods framed under three areas of interest. The first is policy advice and includes municipal level conservation legislation and fisheries protection. The second is promotion of innovative protective measures like using green-gray breakwaters in mangrove rehabilitation and coastal protection, community-led mangrove rehabilitation and community-based mangrove and beach forest nurseries, community-managed seaweed farming with a savings component, and community-based ecotourism management. The third area is awareness raising and knowledge management which includes mangrove and beach forest management training and citizen science.

Nature-based interventions have been implemented at the barangay (village) and municipal scales in the Western Visayas provinces of Aklan, Iloilo, Negros Occidental, and Guimaras.

Governance

Commissioned by the German Federal Ministry for the Environment, Nature Conservation and Nuclear Safety (BMU), ProCoast is part of the Coastal and Marine Ecosystems Management Program (CMEMP) under the Department of Environment and Natural Resources. Its goals are in line with the Biodiversity Management Bureau, who administers a 10-year national program. The project is also supported by the Zoological Society of London (ZSL) and engages the selected local government units, private stakeholders, academia, civil society groups.

Outcome

The project seeks to significantly enhance Marine Protected Areas (MPA) to increase biodiversity for coastal protection and boost the resilience of these ecosystems to the impacts of climate change (Sprong, et al. 2019). The project increases the promotion and dissemination of innovative practices for sustainable use and conservation of coastal resources. It fosters integrative management practices at macro (policy level), meso (support organizations and LGUs), and micro (communities in affected coastal zones) - to protect flora and fauna. Finally, through active involvement of local residents in the political process, the project seeks to encourage greater ownership of the implemented measures. The project champions a positive social environment by promoting social cohesion within local communities.

Writing: Pamela Cajilig
Review: Carole Alazki
Copy Editor: Veronica Olivotto
Design: Phuong Anh Bui

References:

1. Deutsche Gesellschaft für Internationale Zusammenarbeit (GIZ). n.d. 'Protecting Coastal Areas Vulnerable to Typhoons against the Impacts of Climate Change'. Government. Deutsche Gesellschaft Für Internationale Zusammenarbeit (GIZ). Accessed 13 December 2023. https://www.giz.de/en/worldwide/74611.html

2. Oscar M Lopez Center. 2023. 'Sea Level Rise Project – OML Center'. Civic organisation website. OML Center (blog). 2023. https://www.omlopezcenter.org/our-work/sea-level-rise/

3. Sprong, Franca, Terrence Dacles, Raquel Capio, and Rhodora May Sumaray-Raras. 2019. 'Sustainable Protection of Coastal Ecosystems and Communities from the Impacts of Climate Change'. Deutsche Gesellschaft für Internationale Zusammenarbeit (GIZ). https://www.giz.de/en/downloads/20190916_ProCoast_Factsheet.pdf

Photos by Arnel Murga.

AS_12 Conservation of Subak Systems in Bali

Context Overview

NBS TYPOLOGY	ECOSYSTEM TYPE
Protection & conservation; Restoration	Wetlands; Agriculture & food spaces
IMPACT SCALE	**URBAN TYPOLOGY**
Greater urban region	Peri-urban
SOCIAL CHALLENGES ADDRESSED	**CO-BENEFITS**
Food security; Economic vulnerability; Soil quality	Increased biodiversity and their functions; Reinforcement of socio-ecological knowledge; Water security

The *subak* agricultural systems represent a unique indigenous nature-based solution that contributes to food security while providing ecosystem services, habitat for biodiversity and protecting wetlands from urban expansion (Risna et al., 2022). These systems can be found in Bali, a tropical island in Indonesia housing more than 4 million people that has become the center stage for a booming tourism industry. Rice cultivation has cultural and historic significance in the island, taking the form of flat or terraced fields interspersed with low-density built-up areas.

Some rice paddies in Bali are still managed under the millennium-old subak system, which consists of a farmer organization that collectively decides on the irrigation schedule and general resource allocation for agricultural management, following traditional values and local wisdom (Sedana et al., 2021). These systems take in consideration local and indigenous knowledge systems and operate beyond any administrative or political jurisdiction. Subaks have been recognized by the Unesco World Heritage as a manifestation of the Tri Hita Karana philosophy, which brings together the spiritual, human, and natural worlds (UNESCO, 2023).

Introduction to the case study

With the rise of tourism, rapid land use change, and a shift away from agriculture from the young generation, many rice paddies in Bali are no longer managed under the subak system. This has multiple implications for i) water and land management, with land use replacing rice paddies being more water intensive (e.g., hotels, laundromats, etc.) and increasing pollution; ii) ecological integrity, with the lack of synchronization of irrigation schedules promoting pest invasion; and iii) governance, with the rise in conflicts associated with losses of income and water management issues.

Rice plantations have cultural and economic significance to communities living in the urban peripheries of large cities in Southeast Asia.

Nature-based Solutions characteristics

Conservation of the subak systems in Bali can be considered a nature-based solution as it helps preserve the rice paddies and the wetland systems where they are embedded (Risna et al., 2022). This practice maintains key ecosystem services such as sediment retention, water quality regulation and alleviation of flood peaks to the surrounding areas. It also prevents the land from being developed for tourism or residential use, and promotes the sustainable management of the landscape. Although rice paddies are heavily managed, they can be managed sustainably by maintaining solid fertility while following agricultural best practices (by controlling water use, fertilizer or pesticide reliance, etc.)

Governance

The governance of subaks is not tied to administrative or political institutions, but rather follows ancestral watershed-based boundaries. However, the Bali Province and Regency governments provide support for the subak system, for example by encouraging the use of local agricultural products in hotels or restaurants, providing financial support, or re-registering subaks (Traditional Society Development Service of Bali Province).

Outcome

The preservation of the subak systems in Bali have several ecological and societal benefits: i) promoting sustainable water management with incentives to conserve water; ii) regulating cropping patterns, with positive effects on pest controls; iii) generating income for the farmers and the tourism industry related to rice fields; and v) promoting the outstanding universal values embedded in the Tri Hita Karana philosophy, including a harmonious relationship between the Balinese people and their environment through the support and practice of rituals that emphasize our dependencies on healthy ecosystems.

Writing: Perrine Hamel and Erich Wolff
Review: Carole Alazki
Copy Editor: Veronica Olivotto
Design: Phuong Anh Bui

References:

1. Sedana et al., (2021) Roles of local farmers' organization in supporting food security: case of Subak in Bali, Indonesia. Available at: https://doi.org/10.1088/1755-1315/911/1/012083 (Last Accessed: Jul 31, 2025)

2. Risna et al. (2022) Subak, a Nature-based Solutions evidence from Indonesia OP Conf. Ser.: Earth Environ. Sci. 959012030 Available at: https://iopscience.iop.org/article/10.1088/1755-1315/959/1/012030/pdf (Last Accessed: Jul 31, 2025)

3. UNESCO (2023) Cultural Landscape of Bali Province: the Subak System as a Manifestation of the Tri Hita Karana Philosophy (Indonesia) (C 1194rev) Available at: https://whc.unesco.org/en/decisions/8146 (Last Accessed: Jul 31, 2025)

Photo by Erich Wolff.

AS_13 Urban Agriculture at Solar-powered Kochi International Airport

Context Overview

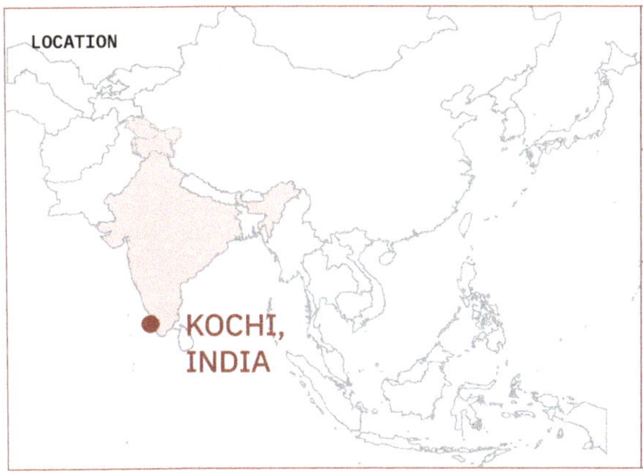

NBS TYPOLOGY	ECOSYSTEM TYPE
Constructed & novel ecosystems	Agriculture & food spaces
IMPACT SCALE	**URBAN TYPOLOGY**
Building	Urban core
SOCIAL CHALLENGES ADDRESSED	**CO-BENEFITS**
Food security; Social inclusion	Green jobs and economic benefits; Carbon sequestration; Urban cooling

Positioned approximately 25 km northeast of Kochi city at sea level, Cochin International Airport (CIA) lies in a tropical monsoon climate with an average temperature of 30°C and substantial rainfall during the southwest monsoon season from June to September. These climatic conditions support lush vegetation and influence the regional landscape. Against this ecological backdrop, the airport—operated by Cochin International Airport Limited (CIAL)—has combined solar energy generation with organic farming beneath and between solar panels. This approach addresses multiple challenges including emissions reduction, food security, and efficient land use.

By leveraging the local ecosystem to support both renewable energy and agriculture, the initiative exemplifies how NbS can generate wide-ranging co-benefits such as biodiversity enhancement, local employment, and reduced food miles, while setting a global benchmark for sustainable infrastructure in the Global South.

Introduction to the case study

This case illustrates how nature-based solutions (NbS) have been innovatively integrated into large-scale infrastructure through the example of CIA in Kerala, India. In a strategic effort to reduce costs and decrease carbon emissions, CIAL invested in renewable energy through a solar photovoltaic (solar PV) system in the early 2000s . The initiative installed a 100 KWp solar PV power plant, making this the first grid-connected solar PV system in Kerala. Building upon the success of this initial project, CIAL furthered its commitment by implementing a 1 MegaWatt peak (MWp) solar PV power plant. Since their inception, these solar facilities have effectively mitigated over 550 metric tonnes of CO2 emissions, significantly contributing to CIAL's environmental conservation initiatives (Koduganti et al., 2023).

Despite receiving acclaim from various national and international entities, CIAL was criticized for the land usage, asserting that the valuable land should have been utilized for addressing hunger in India. This criticism prompted officials to act. To optimize land usage and enhance green spaces, CIAL established organic vegetable gardens on approximately three acres of land in and around the periphery of the solar PV plant (Koduganti et al., 2023). This dual-use of land conserves space and resources, turning otherwise unused areas into productive farmland.

Nature-based Solutions characteristics

CIAL has embraced a multifaceted NbS approach incorporating the principles of sustainable development and greenhouse gas emissions reduction. As a part of its commitment to developing nature-based solutions, the facility has integrated 46,150 solar panels across 45 acres. These panels include 400 polycrystalline modules with a capacity of 250Wp each, along with five 20 kW capacity Refu-sol string inverters. Additionally, the facility incorporates a small hydroelectric project that has generated approximately 7.5 million units of electricity to date. Since 2016, a distinct unit within CIAL has been engaged in cultivating vegetables amidst the solar panels, with a spacing of 1.5 meters. The cultivation focused on creepers and shrub varieties of vegetables such as okra, long beans, pumpkin, ash gourd, melons, and ginger (CIAL, 2018). The airport's sustainability model operates at a neighborhood scale but has global implications. It utilizes the tropical ecosystem to support its solar energy and organic farming initiatives.

Governance

Established in 1999, Cochin International Airport is a greenfield airport developed through a public-private partnership. The solar power project at the airport, which required an investment of INR 62 crore, recouped its costs within six years, proving its financial sustainability. The project was led by Cochin International Airport Limited (CIAL) with notable contributions from the Kerala State Government and Non-Resident Indians (NRIs) from Kerala who are stakeholders. The vegetable gardens' daily upkeep—such as watering, weeding, and seasonal harvesting—is managed by a team of 12-14 laborers. This initiative is backed by Kerala's regulatory framework for electricity and environmental policies and supports India's commitment to achieving net-zero emissions by 2070

Outcome

This approach yielded significant results in reducing emissions, generating employment, and producing fresh, organic produce (refer to Figure 14). Some airport officials and laborers working on the vegetable farms reported a reduction in ambient heat as a result of the cultivation. Acknowledging this forward-thinking strategy, the airport was honored with the prestigious Champion of the Earth award in 2018, the highest environmental accolade bestowed by the United Nations.

While Cochin has set the precedent, the remainder of India is on the brink of tapping into the immense potential of solar energy. However, obstacles remain: the absence of a stable electricity grid capable of handling anticipated increases in power capacity and the lack of established standards for solar power generation and usage. Additionally, communities tend to be mindful of costs, and when choosing rooftop power systems, they might opt for cheaper equipment, which could lead to a compromise in quality.

Cultivating vegetables was not initially within the objectives of CIAL. Harvesting certain vegetables, like cluster beans, posed a serious challenge. The relentless growth of weeds presented another difficulty that laborers found hard to address. Recognizing that daily harvesting was essential for sales at the airport stalls, CIAL experimented with cultivating vegetables with longer growing seasons, such as ginger and turmeric (Menon, 2015; Express News Service, 2016).

In summary, the transition to solar energy, the establishment of organic vegetable farms, and the commitment to optimizing land use have demonstrated that large-scale infrastructure projects can attain full operational and 'power neutral' status through alternative energy sources, setting a global precedent.

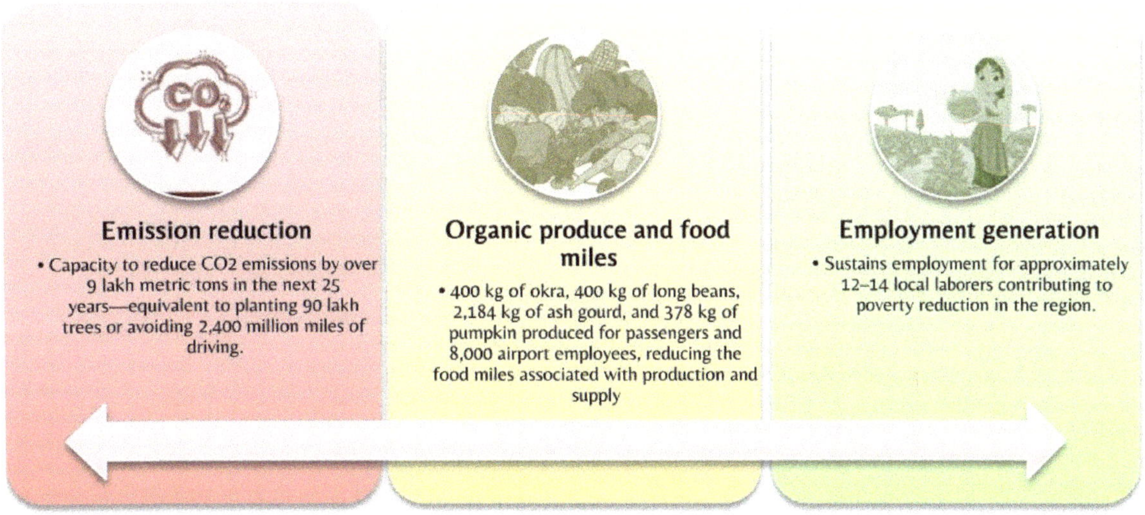

Figure 14: Outcomes of NbS at Cochin International Airport.

Writing: Maitreyi Koduganti & Sheetal Patil (Indian Institute for Human Settlements, India)
Review: Carole Alazki
Copy Editor: Veronica Olivotto
Design: Phuong Anh Bui

References:

1. CIAL (n.d.). Investor Info. Cochin International Airport Ltd. https://web.archive.org/web/20101215124902/http://cochin-airport.in/contents/viewcontent.aspx?linkIdLvl2=55&linkId=55
2. CIAL. (2016). Organic farming. Cochin International Airport Limited. Retrieved from https://www.cialinfra.in/Projects/ORGANICFARMING
3. CIAL. (2018, July 27). CIAL wins UN's highest environmental honour. The Hindu. Retrieved from https://www.thehindu.com/news/national/kerala/cial-wins-uns-highest-environmental-honour/article24524222.ece
4. Express News Service. (2016). Next: Organic farming at CIAL. The New Indian Express. Retrieved from https://www.newindianexpress.com/cities/kochi/2016/mar/05/Next-Organic-Farming-at-CIAL-899892.html
5. Jayan, P. (2016, June 1). Cochin International Airport - World's first solar-powered airport. ICAO ENVIRONMENTAL REPORT, 177. https://www.icao.int/environmental-protection/Documents/EnvironmentalReports/2016/ENVReport2016_pg177-177.pdf
6. Koduganti, M., Patil, S., & Singh, C. (2023). Suriya Prakashathil Krishi: The Story Of Cochin Airport (pp. 50-53). Indian Institute for Human Settlements. https://doi.org/9788195847310
7. Menon, S. (2015, October 9). How is the world's first solar powered airport faring? BBC. https://www.bbc.com/news/world-asia-india-34421419

Diagram by Maitreyi Koduganti & Sheetal Patil

AS_14 Puey Ungphakorn Centenary Hall Green Roof

Context Overview

LOCATION

NBS TYPOLOGY	ECOSYSTEM TYPE
Constructed & novel ecosystems	Roofs & walls; Green spaces (gen.)
IMPACT SCALE	**URBAN TYPOLOGY**
Building	Peri-urban
SOCIAL CHALLENGES ADDRESSED	**CO-BENEFITS**
Flooding; Heat	Community engagement & social cohesion; Reinforcement of socio-ecological knowledge; Food security

The Puey Ungphakorn Centenary Hall is a large green roof that demonstrates the ecological, educational and cultural benefits of integrating nature-based solutions into architecture. The roof is a landmark of the Rangsit campus of Thammasat University in the Bangkok Metropolitan Region, Thailand. Located at the urban expansion border of the largest city in Thailand, Rangsit is losing low-intensity agricultural fields and vegetation due to fast-paced urbanization. The region experiences a tropical monsoon climate and faces urban climate challenges such as flooding and the urban heat island effect (Loc et al., 2023), aggravated by land use change and the degradation of ecosystems.

Introduction to the case study

The city of Bangkok is highly vulnerable to floods due to the expansion of impervious surfaces and an over dependence on gray infrastructure (Rauf et al., 2024). The catastrophic floods in 2011 have revealed a fragmented governance system and the need for more investments in flood risk reduction in the city, particularly in upstream areas such as Rangsit. The loss of green public spaces and the quick urbanization experienced in Bangkok are expected to aggravate current concerns with floods and urban heat islands (Ramsay et al., 2023), disproportionately affecting communities in historically marginalized areas. The Puey Ungphakorn Centenary Hall is intended to provide a public space for educational and cultural activities that is accessible to nearby residents and the academic community.

Vegetation growing on the roof of the Puey Ungphakorn Centenary Hall, at Thammasat University

Nature-based Solutions characteristics

The 22,000-m^2 Puey Ungphakorn Centenary Hall green roof is an example of innovative architecture using nature to address societal issues. Its design leverages Water Sensitive Urban Design techniques to direct stormwater to gardens that were designed to retain sediments, improve water quality and support vegetation. The roof comprises a 7000 m2 urban farm, Asia's largest organic rooftop farm, which grows more than 40 edible species including rice, herbs, and fruit trees.

Governance

The Puey Ungphakorn Centenary Hall green roof was completed in 2019 for an estimated cost of 944,000,000 THB (USD 26,000,000). It was commissioned and is managed by Thammasat University, a public institution. The project is expected to benefit from the involvement of students, university staff and surrounding residents in educational activities at the rooftop farm, which will foster stewardship and community bonding.

Outcome

The green roof was designed to serve several purposes. First, it addresses flood issues by reducing and slowing down the amount of runoff during rain events. The terrace-inspired design provides more opportunities for runoff to be retained. The green roof may also help reduce the urban heat island effect by reducing heat storage and insulating the building. Importantly, the roof's urban farm addresses food security and serves multiple education purposes for students and the public on sustainable urban farming.

The green roof has been praised by the architectural community for its innovative design that helps address climate challenges such as flooding and urban heat, and provides food, and educational services. Little to no evidence of these outcomes is available publicly.

Writing: Erich Wolff and Perrine Hamel
Review: Carole Alazki
Copy Editor: Veronica Olivotto
Design: Phuong Anh Bui

References:

1. Ramsay, Emma E., Genie M. Fleming, Peter A. Faber, S. Fiona Barker, Rohan Sweeney, Ruzka R. Taruc, Steven L. Chown, and Grant A. Duffy. 2021. "Chronic Heat Stress in Tropical Urban Informal Settlements." iScience, November, 103248. https://doi.org/10.1016/j.isci.2021.103248.

2. Rauf, Hanna A., Erich Wolff, Boonanan Natakun, Wijitbusaba Marome, and Perrine Hamel. 2024. "Aligning Nature-Based Solutions and Housing Policy: A Study on Stakeholders' Perceptions of Nature within Informal Settlement Upgrading Projects." Cities 153 (October):105264. https://doi.org/10.1016/j.cities.2024.105264.

3. Ho Huu Loc, Adel Emadzadeh, Edward Park, Piyanuch Nontikansak, Ravinesh C. Deo,

4. The Great 2011 Thailand flood disaster revisited: Could it have been mitigated by different dam operations based on better weather forecasts?, Environmental Research, Volume 216, Part 2, 2023, 114493, ISSN 0013-9351, https://doi.org/10.1016/j.envres.2022.114493.

AS_15

Putrajaya Wetlands Park

Context Overview

NBS TYPOLOGY Restoration; Constructed & novel ecosystems	ECOSYSTEM TYPE Riparian areas; Green spaces (gen.); Wetlands
IMPACT SCALE Neighbourhood/Community	URBAN TYPOLOGY Urban core
SOCIAL CHALLENGES ADDRESSED Flooding; Water quality; Nature connection	CO-BENEFITS Increased biodiversity and their functions; Recreational benefits

The Putrajaya Wetlands are situated in Putrajaya (Malaysia) and serve as an example of how constructed wetland ecosystems can be integrated into parks within urban environments. The city of Putrajaya itself is a planned administrative and governmental green city, located south of Kuala Lumpur. The wetland is 600 hectares, designed to treat surface runoff from the surrounding area and two main rivers that flow into Putrajaya Lake, Sungai Chua and Sungai Bisa (Urban Nature Atlas 2023, Perbadanan Putrajaya 2001). Rock-filled weirs were built across the five arms of the wetland, partitioning the wetland into the 24 cells. These five arms are interconnected, but vary in their size, depth, plant ecosystems, and the types of pollutants they are intended to manage. Surrounding the wetlands is a park area, with a range of recreational facilities. The region experiences a tropical rainforest climate with wet and dry seasons.

Introduction to the case study

The establishment of the Putrajaya Wetlands between 1997 and 1998 played a significant role in the broader context of relocating the capital of Malaysia from Kuala Lumpur to Putrajaya. This move aimed to not only decongest the capital but also pave the way for the development of a new, environmentally conscious, and modern city. The city was planned as a garden city and smart city, with 38% of the area allocated to green open spaces. The centerpiece of the city was the Putrajaya wetlands (Urban Nature Atlas 2023). More recently the city has also adopted a low carbon green city strategy (Putrajaya Corporation 2012).

Nature-based Solutions characteristics

The Putrajaya Wetlands project employs various nature-based solutions for water quality improvement and flood mitigation. The project restores water quality by natural processes, such as filtration and sedimentation and helps regulate water flow and reduce flood risks. The nature-based solutions used are also expected to protect and enhance the wetland and surrounding terrestrial ecosystems through restoration and conservation as well as mitigating high temperatures associated with surrounding urban fabric. Among the key co-benefits, the park has been credited for creating green spaces for residents and visitors to enjoy nature (Smart Putrajaya 2021). The wetlands encompass a range of ecosystems, including lakes, ponds, marshes, and forests. They also include nature areas supporting biodiversity conservation and multifunctional areas with green open spaces for recreation.

These areas are integrated at a neighborhood and city scale, providing a holistic approach to urban planning and water management.

Governance

In 1995, the Federal Government of Malaysia endorsed a Master Plan for the establishment of the Putrajaya Federal Government Administrative Centre. The administration of this city falls under the jurisdiction of the Putrajaya Corporation. The Perbadanan Putrajaya (PPJ) is a local authority that administers the Federal Territory of Putrajaya, under the Federal Territories Ministry of Malaysia. The PPJ led the Putrajaya wetlands project within a Local Agenda 21 programme and action plan to achieve sustainable development that offers employment opportunities and raises awareness of local stakeholders on the physical, economic, social and environmental implications of the project. The project costs amount to more than 4,000,000 EUR (more than 4,680,000 USD at 2025 rates).

Walkway overlooking the Putrajaya Wetlands (Malaysia)

Outcome

The Putrajaya Wetlands and the surrounding urban environment represent a pioneering example of sustainable urban design, constructed even before the term "nature-based solutions" came into common use. The immense size of these wetlands, recognized as the largest constructed wetlands in the Southern Hemisphere, and their successful three-decade-long operation underscore the commitment and effectiveness of their design. Putrajaya Wetlands stands as a model for the integration of nature-based solutions into urban planning, effectively addressing environmental concerns, improving quality of life, and promoting sustainable development.

Writing: Alex Lechner
Review: Carole Alazki
Copy Editor: Veronica Olivotto
Design: Phuong Anh Bui

References:

1. Perbadanan Putrajaya 2001 Putrajaya Lake Use And Navigation Master Plan And Lake And Wetland Emergency Response Plan Final Report (2nd Revision) August 2001 (2nd Revision) https://www.ppj.gov.my/storage/putrajaya06/269/269.pdf

2. Putrajaya Corporation 2012 Putrajaya Low Carbon Green City Initiative https://www.ppj.gov.my/storage/laporan04/654/654.pdf

3. Smart Putrajaya 2021 Putrajaya Lake And Wetland https://smart.putrajaya.my/project/putrajaya-lake-and-wetland/

4. Urban Nature Atlas 2023 Putrajaya: Pioneer Green City https://una.city/nbs/putrajaya/putrajaya-pioneer-green-city

Photo by Alex Lechner

AS_16

Fukuoka Hall Green Roof

Context Overview

NBS TYPOLOGY Constructed & novel ecosystems	ECOSYSTEM TYPE Roofs & walls
IMPACT SCALE Building	URBAN TYPOLOGY Urban core
SOCIAL CHALLENGES ADDRESSED Physical and mental health; Social wellbeing; Social inclusion	CO-BENEFITS Recreational benefits; Increased biodiversity and their functions; Flood mitigation; Urban cooling

The ACROS Fukuoka Prefectural International Hall Step Garden is a flagship green roof project situated within Fukuoka's financial district, bordering the Naka River. Fukuoka is the most populous area in Kyushu Island and is among the largest cities in Japan. This "Step Garden" is well adapted to the humid subtropical climate of the northern coast of Kyushu and showcases a 14-tier green roof incorporated into the Asian Cross Roads Over the Sea (ACROS) Fukuoka Prefectural International Hall, which was completed in 1995 (Urban Nature Atlas 2021). Together with the surrounding Tenjin Central Park this rooftop garden is recognized as the sole open green space available in the city center.

Introduction to the case study

The Fukuoka Prefecture International Hall Step Garden is a response to the plans to construct a substantial government office building in the central urban area of Fukuoka, which would require the loss of vegetated space in the city (Urban Nature Atlas 2021). The original blueprint threatened to claim half of Tenjin Central, a rare green oasis in this bustling commercial district. In response to public opposition, a compromise was reached by concealing the office building beneath a green roof, thereby safeguarding the entire two-hectare park (Belogolovsky 2020).

Nature-based Solutions characteristics

The Fukuoka Prefectural International Hall Step Garden utilizes nature-based solutionsby implementing green roofs, vertical gardens, and native vegetation. Initially, 37,000 plants from 76 different varieties were planted during construction. Over time, additional planting and seed dispersal by wild birds have increased the garden's biodiversity to 120 varieties and a total of 50,000 plants. While the original intention was primarily around aesthetic and recreational values, the green roof will also have benefits for combating urban heat island effect, enhancing stormwater absorption, and improving overall ecosystem health in the vicinity.

Governance

The original building construction was led by the city administration and currently the ACROS Fukuoka Foundation manages and administers the public sections of the ACROS Fukuoka building (approximately 40% of the total building) on behalf of Fukuoka Prefecture (Urban Nature Atlas 2021).

Outcome

The ACROS Fukuoka Prefectural International Hall, constructed nearly 30 years ago, remains a lasting cultural symbol and the city's most frequented landmark. Designed by Emilio Ambasz, a prominent architect and industrial designer, this early example of green architecture continues to be a widely cited case study. It has stood the test of time and showcases the profound environmental benefits of a sustainable project for both people and the environment. The rooftop garden is well regarded by the residents as it resembles a lush cascade of greenery, flowing down the stepped sides of the building. This feature has cultural and symbolic value, earning it the nickname "ACROS Mountain" due to its natural mountain-like appearance. This expansive rooftop green area spans 5,400 square meters, spanning from the 2nd floor to the 14th floor.

Writing: Alex Lechner
Review: Carole Alazki
Copy Editor: Veronica Olivotto
Design: Phuong Anh Bui

References:

1. Urban Nature Atlas (2021) ACROS Fukuoka Prefectural International Hall Step Garden. Urban Nature Atlas. https://una.city/nbs/fukuoka/acros-fukuoka-prefectural-international-hall-step-garden

2. Belogolovsky, V. (2020) ACROS Fukuoka Prefectural International Hall by Emilio Ambasz turns 25. Stir World

3. https://www.stirworld.com/think-columns-acros-fukuoka-prefectural-international-hall-by-emilio-ambasz-turns-25

AS_17 From Farm to Classroom: Bengaluru's Ragi Project

Context Overview

LOCATION

NBS TYPOLOGY	ECOSYSTEM TYPE
Constructed & novel ecosystems; Restoration	Agriculture & food spaces
IMPACT SCALE	**URBAN TYPOLOGY**
Site	Urban core
SOCIAL CHALLENGES ADDRESSED	**CO-BENEFITS**
Nature connection; Land degradation	Reinforcement of socio-ecological knowledge; Community engagement & social cohesion; Food security

Bengaluru is located in the southeastern part of the state of Karnataka, in southern India. Situated in a peri-urban area undergoing rapid land-use change, the Ragi Project at the Poorna Learning Centre in Bengaluru, responds to the growing disconnect between urban communities and nature by using school-based farming as a tool for environmental education, sustainability, and cultural revival. Nicknamed the "Garden City of India", Bengaluru has long been known for its parks and tree-lined avenues—though rapid urbanization has put pressure on green cover.

Introduction to the case study

The Ragi Project is a powerful example of how nature-based solutions can be integrated into educational settings. The primary issues addressed by the Ragi Project included the loss of agricultural land to urban development and the disconnection of urban populations, especially children, from food production processes. The project aimed to engage students in hands-on farming activities, fostering a deeper understanding of sustainable agriculture and local food systems. Additional challenges included promoting healthier eating habits and reviving traditional crops like ragi (finger millet), which are well-suited to the local climate and culture.

Activities on the farm.

Nature-based Solutions characteristics

The Ragi Project emphasized both conservation and development by restoring a portion of land for farming and integrating food education into the school curriculum. This innovative approach combined practical farming experience with classroom learning, emphasizing sustainable agricultural practices. Operating within an agricultural ecosystem, the project utilized organic farming techniques, such as composting and preparing natural fertilizers like jeevamrutha, to enhance crop cultivation and soil health. It also employed biomimicry principles by incorporating traditional farming methods and local ecological knowledge. Implemented on a small scale, the project involved a 0.5-acre plot of land near the school and engaged students from various grades in regular farming activities, seamlessly integrating these experiences into their educational curriculum (Ravi & Patil, 2023).

Governance

The Ragi Project involved a diverse group of actors, including students, teachers, parents, and local community members. Notably, the school's support staff, with significant contributions from Radha, who generously offered her land for farming, played a crucial role in the project's success. The project's financial backbone was primarily the school community, with additional resources garnered through active community involvement and contributions. Operating within the institutional framework of the Poorna Learning Centre, the project was bolstered by policies and practices that promoted sustainable and ecologically conscious initiatives, ensuring its integration into the school's curriculum and daily activities (Ravi & Patil, 2023).

Outcome

Spanning from 2017 to 2019, the project saw a range of activities and outcomes documented over this period . It provided educational impact by offering hands-on learning experiences that bridged the gap between students and the natural world, as well as agriculture. This approach facilitated interdisciplinary learning across various subjects, including science, geography, history, and mathematics, through practical farming activities. Additionally, the project played a crucial role in community building by fostering a sense of collaboration among students, teachers, and parents, and promoting collective problem-solving. It also enhanced sustainability awareness, equipping students with a deeper understanding of sustainable agriculture and food systems and encouraging environmentally conscious behaviors and choices (Kilpadi, 2017). Furthermore, the project contributed to cultural revival by rekindling interest in traditional crops like ragi, which are both nutritious and well-suited to the local environment, while reconnecting students with cultural practices and food traditions.

The journey of developing, implementing, and sustaining Ragi Project Ragi Project was not without its hurdles. There were significant resource constraints, such as finding a reliable water source and managing the farm during school holidays. Additionally, acquiring the necessary farming skills and expertise was a steep learning curve. Teachers had to get creative and flexible to integrate farming activities into the school curriculum, often reorganizing the timetable to fit in farm visits and related tasks. Keeping the project alive and thriving beyond the initial excitement was another big challenge, requiring continuous commitment from the entire school community and adapting to changing seasons, pest attacks, and specific growth periods of the crops. Despite these challenges, the collective effort and dedication of everyone involved kept the project moving forward.

The Ragi Project successfully illustrated how urban school farming can meet multiple educational and environmental objectives. By involving students in hands-on farming activities and weaving these experiences into the curriculum, the project deepened their understanding of sustainable agriculture, local food systems, and cultural traditions. Despite facing challenges, the project's holistic approach and strong community involvement underscored the potential for similar initiatives to advance sustainability and education in urban environments. Ultimately, the Ragi Project stands as a valuable model for integrating food education into school curricula, promoting sustainable agricultural practices, and fostering a sense of community and environmental stewardship among students.

Educational Impact

provided hands-on learning experiences that connected students with nature and agriculture

Community Building

fostered a sense of community among students, teachers, and parents, while encouraging collaboration and collective problem-solving.

Awareness

a deeper understanding of sustainable agriculture and food systems, promoting environmentally conscious behaviors and choices.

Cultural Revival

revived interest in traditional crops like ragi, while reconnecting students with cultural practices and food traditions.

Figure 15: Outcomes of the Ragi Project.

Writing: Maitreyi Koduganti & Sheetal Patil (Indian Institute for Human Settlements, India)
Review: Carole Alazki
Copy Editor: Veronica Olivotto
Design: Phuong Anh Bui

References:

1. Kilpadi, M., (2017). The Ragi Project: How one school is feeding itself from within. EarthaMag. Available at: https://www.earthamag.org/stories/stories/2017/12/11/the-ragi-project-how-one-school-is-feeding-itself-from-within (Last Accessed June 2025)

2. Ravi, R., and Patil, P. V. (2023). Learning through Food: The Ragi Project in Bengaluru. In Sowing Sustainable Cities. Indian Institute for Human Settlements. Available at: https://doi.org/10.24943/SSC12.2023 (Last Accessed June 2025)

Photo by Roshni Ravi
Diagram by Maitreyi Koduganti

AS_18 Making Space for Fauna in Bengaluru

Context Overview

NBS TYPOLOGY Constructed & novel ecosystems; Restoration	ECOSYSTEM TYPE Agriculture & food spaces; Lakes & ponds; Wetlands
IMPACT SCALE Site	URBAN TYPOLOGY Urban core
SOCIAL CHALLENGES ADDRESSED Land degradation; Water quality; Biodiversity loss; Heat	CO-BENEFITS Reinforcement of socio-ecological knowl-edge; Food security; Green jobs and econom-ic benefits

Bengaluru lies on the Deccan Plateau about 920 meters above sea level, with semi-tropical temperatures that hover between 20 and 32 °C throughout the year. India's technology capital has grown from roughly three-quarters of a million residents in 1950 to more than fourteen million in 2024, making it one of the world's fastest-expanding metropolitan areas. Rapid urbanization has consumed almost all of the city's historic tree canopy and wetlands: satellite analysis by the Indian Institute of Science shows built-up land rising from about 2% of the municipal area in 1973 to nearly half by 2016, while green cover has fallen from an estimated 68% in the early 1970s to 6% today (Energy & Wetlands Research Group, 2025).

The city provides barely 2.2 square meters of public open space per person, far short of the ten to twelve square meters recommended by India's Ministry of Urban Development, and tree counts average only one for every seven residents rather than the eight per capita needed to offset basic carbon emissions (Ramachandra et al.n.d). The loss of vegetation has intensified urban heat-island effects, degraded air and water quality, lowered groundwater recharge, and eroded habitat for the birds, bats, insects, and small mammals that once thrived in Bengaluru's "garden city" landscape. Temperatures in streets devoid of shade run as much as five degrees Celsius higher than in the few remaining parks, and surface readings on unshaded roads can exceed nearby vegetated areas by more than twenty degrees.

Introduction to the case study

The Bengaluru pilot seeks to restore three essential ecosystem services that have eroded under rapid urbanization: pollination for household food production, biological control of rodents and insect pests, and in-lake purification of nutrient-laden water. To meet these goals, the project combines habitat devices such as bee hotels and owl boxes with community partnerships that bring traditional rodent-catching skills to campus grounds, and it supplements hard infrastructure with floating vegetated islands that strip nutrients from stagnant ponds. By linking household-scale and neighborhood-scale interventions, Indian Institute for Human Settlements (IIHS) aims to show city agencies, resident-welfare associations, and CSR funders how targeted, replicable actions can enhance biodiversity, reduce chemical dependence, and improve the quality of food and water in Bengaluru's dense urban core.

Nature-based Solutions characteristics

The Bengaluru pilot links three habitat-based interventions to specific urban issues. Pollination in rooftop and courtyard food plots is boosted with simple wooden-and-bamboo bee hotels installed among vegetable beds; these shelters provide nesting cavities for solitary bees and wasps, which in turn improve fruit set and crop yields while strengthening local pollinator populations. For biological pest control, the team mounts owl nest boxes high in mature trees and quiet building corners, encouraging owls—and, where suitable, bats and small insectivorous birds—to roost and breed near human activity; their natural diets of rats, mice, cockroaches and beetles reduce reliance on chemical pesticides. On the IIHS Kengeri campus, shelter provisioning is reinforced by a ten-day rodent drive led by members of the Indigenous Irula community, whose traditional live-capture methods removed 150 rodents without poisons and demonstrated a cultural pathway for ongoing pest management. Water-quality concerns are addressed with two to three floating vegetated islands anchored in campus ponds and the Tambraparni basin. These are lightweight frames topped with coir or mesh mats planted with native sedges, grasses and floating aquatics that take up excess nitrogen, phosphorus and heavy metals. The plants are harvested periodically so the captured pollutants are physically removed from the water body. Together these low-cost, locally fabricated devices restore pollination, predation and nutrient-filtration functions to fragmented urban ecosystems while remaining simple enough for neighborhood groups, resident-welfare associations and CSR sponsors to replicate across the city.

Governance

Urban nature-based solutions in Bengaluru operate at many scales from single-house gardens and apartment terraces to institutional campuses and public lakes, and their governance must reflect this diversity. Outreach begins with social media posts, exhibitions, and workshops that teach residents how to obtain pollinator-friendly plants or construct bee hotels, while resident-welfare associations manage budgets and approvals for entire complexes. Low-income areas often need subsidized kits that are provided through corporate-social-responsibility funds and groups such as the Bengaluru Sustainability Forum. Larger projects that restore parks, lakes, or urban forests require agreements with agencies like the Bruhat Bengaluru Mahanagara Palike, the Horticulture Department, and the Forest Department, and they are increasingly incorporated into initiatives such as the Bengaluru Climate Action and Resilience Plan. Companies seeking environmental offsets increasingly collaborate with NGOs and research institutes, and the expanding use of peri-urban Other Effective Area-Based Conservation Measures (OECMs) provides campuses with a formal framework and fresh CSR funding for biodiversity-friendly management, thereby advancing India toward its Kunming-Montreal goal of conserving 30 percent of its land by 2030. Additional finance may come from national adaptation funds, the Green India program, and the emerging green-credits policy, which could support urban nature-based solutions where offset rules permit.

Outcome

Most owl, bat, and bee boxes on green campuses and suburban rooftops attracted occupants within the first season, confirming that artificial shelters can quickly boost urban wildlife when surrounding habitat is adequate. In the city's most densely built quarters, however, the same boxes remained empty for months, a clear sign that basic foraging and nesting habitat must be restored before such devices can work. The floating vegetated islands rooted within weeks, and water-quality samples taken downstream of the rafts already show modest declines in nutrient and turbidity levels. Their chief drawback is upkeep: fast-growing sedges and grasses must be harvested every few months to remove absorbed pollutants and keep the islands buoyant, and replacement plantings are needed where growth rates lag.

The interventions deliver several co-benefits. Bee hotels raise pollination rates in rooftop gardens and community plots, owl and bat boxes cut reliance on chemical rodenticides and insecticides, and the islands add greenery while filtering lake water. Yet each device demands regular inspection. Boxes can be taken over by ants, squirrels, or non-target wasps and sometimes require repair or relocation, while unmanaged plant rafts can become overgrown and lose efficiency. Sustained monitoring and community stewardship therefore remain the decisive factors for long-term success.

Writing: Ravi Jambhekar, Jagdish Krishnaswamy (Indian Institute for Human Settlements, India)
Review: Carole Alazki
Copy Editor: Veronica Olivotto
Design: Phuong Anh Bui

References:

1. Seto, K.C. and Ramankutty, N., 2016. Hidden linkages between urbanization and food systems. Science, 352(6288), pp.943-945.Environmental Sustainability, 2(3), pp.127-128.

2. Wenzel, A., Grass, I., Belavadi, V.V. and Tscharntke, T., 2020. How urbanization is driving pollinator diversity and pollination–A systematic review. Biological Conservation, 241, p.108321.

3. Mukherjee, R., Deb, R., & Devy, S. M. (2019). Diversity matters: Effects of density compensation in pollination service during rainfall shift. Ecology and evolution, 9(17), 9701-9711.

4. Ramachandra, T. V., Bharath, H. A., & Sowmyashree, M. V. (n.d.). Green Space in Bengaluru. Energy & Wetlands Research Group, Indian Institute of Science. https://wgbis.ces.iisc.ac.in/energy/water/paper/Green_Space_in_Bengaluru/index.html

5. Energy & Wetlands Research Group. (n.d.). Bengaluru Urban Information System (BUiS). Indian Institute of Science. https://wgbis.ces.iisc.ac.in/sdss/BUiS/

AS_19

Tashkent Green City initiative for Air Quality and Heat Mitigation

Context Overview

NBS TYPOLOGY	ECOSYSTEM TYPE
Restoration	Forests & trees; Green spaces (gen.)
IMPACT SCALE	**URBAN TYPOLOGY**
City	Metropolitan
SOCIAL CHALLENGES ADDRESSED	**CO-BENEFITS**
Heat; Air quality; Nature connection	Flood mitigation; Recreational benefits; Human health and wellbeing

Uzbekistan, a landlocked country in Central Asia, continues to grapple with the legacy of intensive industrialization and urban growth during the Soviet era—marked by water overuse, soil degradation, and deteriorating air quality. These problems are now intensified by climate change impacts, including rising temperatures and the accelerated melting of mountain glaciers (Veckalne & Tambovceva, 2023; Ministry of Ecology, Environmental Protection, and Climate Change, 2023).

Tashkent, home to 2.9 million residents, exemplifies these pressures. The city has experienced significant loss of green spaces and agricultural land and rising levels of air pollution, and the emergence of urban heat islands (Aslanov et al., 2021). These conditions are further strained by the region's Mediterranean climate, with long and intensely hot summers (Mukhamedjanov, Isamukhamedova, & Tang, 2024). In response, the city is embracing large-scale greening interventions, such as the Yashil Makon initiative and the Tashkent Green City program. These efforts aim to restore ecological balance, expand urban tree cover, rehabilitate degraded land, and foster healthier, more resilient urban environments—demonstrating the critical role of NbS in driving sustainable transformation in Central Asian cities.

Introduction to the case study

Tashkent Green City initiative illustrates how nature-based solutions are being strategically implemented to address the compounded environmental challenges of rapid urbanization and climate stress. Tashkent faces mounting environmental challenges exacerbated by climate change. Rising temperatures and the accelerated melting of mountain glaciers, vital for the region's water supply, compound existing vulnerabilities. While forests cover only 7.7% of Uzbekistan's territory—far below the global average—they play a critical role in combating desertification and safeguarding biodiversity (Ministry of Ecology, Environmental Protection and Climate Change, 2023).

Tashkent's rapid urbanization and industrialization have led to significant environmental degradation. The city's average temperatures have risen sharply, influenced by climate shifts and the ecological fallout from the Aral Sea crisis (Kholmatjanov et al., 2020). Air pollution frequently exceeds health standards, with high levels of PM 2.5 driven by industrial emissions, vehicular pollution, and dust.

Moreover, unsustainable waste management practices exacerbate environmental risks. Tashkent produces over 12,000 tons of municipal solid waste daily, with insufficient recycling infrastructure leading to water and soil contamination from industrial chemicals and textile dyes (Asian Development Bank, 2021).

These challenges highlight the urgent need for transformative interventions like the Tashkent Green City initiative, which integrates nature-based solutions to address urban sustainability, enhance climate resilience, and improve overall livability.

Nature-based Solutions characteristics

In Tashkent, the Green City initiative seeks to enhance urban green spaces, improve air quality, and reduce urban heat islands, contributing to both environmental sustainability and biodiversity. Key actions under the initiative include:

- Creation of 4,000 hectares of green zones: These spaces enhance biodiversity, provide recreational areas, and contribute to better air quality.
- Integration of energy-efficient and sustainable practices: Urban planning efforts incorporate energy-saving technologies and green infrastructure to promote long-term resilience.
- Tree-planting campaigns: Public participation in greening efforts is encouraged, fostering environmental stewardship among residents.

Governance

Uzbekistan is increasingly adopting nature-based solutions as part of its climate strategy, including in its Nationally Determined Contributions (NDCs). Notable initiatives like the Yashil Makon (Green Space) project, launched in 2021 with UNECE and UNDP support, aim to plant 200 million trees annually to combat desertification (UNECE, 2022).

The Ministry of Ecology, Environmental Protection, and Climate Change leads nature-based solutions (NbS) implementation as done in Tashkent, with support of UNDP and IUCN. Local NGOs and community groups play a crucial role in the execution and maintenance of NbS projects (IUCN, 2023).

The Uzbek government also allocates national budget funds for NbS, guided by strategic national policies and frameworks. For instance, Uzbekistan's national biodiversity policy aligns with global frameworks such as the Kunming-Montreal Global Biodiversity Framework, reinforcing NbS within the broader environmental strategy (UNDP, 2023). The Strategy for Developing the New Uzbekistan (2022-2026) includes NbS components focusing on green cover expansion and sustainable urban development (IUCN, 2023).

Furthermore, The Asian Development Bank's GREEN project (Growth, Resilience, Energy Efficiency, and Nature-Based Solutions) supports Uzbekistan's initiatives, in particular Tashkent Green City, by providing financial and technical assistance to enhance urban livability through sustainable practices (Asian Development Bank, 2023). Uzbekistan collaborates with international organizations like the Global Green Growth Institute (GGGI) to integrate green growth strategies into national policies. These partnerships help Uzbekistan access expertise and resources needed to implement effective NbS projects (Global Green Growth Institute, 2021).

Outcome

Tashkent has made significant progress in adopting nature-based solutions (NbS) to tackle pressing environmental issues, stemming from rapid urbanization and industrialization in the Soviet era (Veckalne and Tambovceva, 2023). Key initiatives like the Tashkent Green City aim to expand urban green spaces, improve air quality, and promote sustainability. These efforts are complemented by afforestation in the Aral Sea region and the development of drought-resistant plantations to combat desertification.

The initiative has already demonstrated several co-benefits, including:

- Improved Air Quality: New green zones act as carbon sinks and reduce particulate matter (PM 2.5) levels in high-pollution areas.
- Heat Mitigation: The expanded vegetation cover has lowered urban temperatures in pilot areas, improving thermal comfort for residents (Kholmatjanov et al., 2020).
- Recreational Spaces: New parks provide much-needed spaces for leisure, promoting mental and physical well-being.

Governance frameworks, led by the Ministry of Ecology, Environmental Protection, and Climate Change, supported by international partners like UNDP and IUCN, play a pivotal role in managing these initiatives (IUCN, 2023a). However, challenges persist, including the need for sustained financial investment, addressing rapid urbanization, and fostering greater public participation. The success of NbS in Tashkent depends on ongoing support from both national and international stakeholders, as well as the integration of NbS into broader environmental and socio-economic policies.

Writing: Divyanshi Vyas (Indian Institute for Human Settlements, India)
Review: Carole Alazki
Copy Editor: Veronica Olivotto
Design: Phuong Anh Bui

References:

1. Kholmatjanov, B.M. et al. (2020). 'Analysis of temperature change in Uzbekistan and the regional atmospheric circulation of Middle Asia during 1961–2016', Climate, 8(9), p. 101. doi:10.3390/cli8090101.
2. United Nations Economic Commission for Europe Information Unit. (2022). UNECE and UNDP support Uzbekistan's Yashil Makon/Green Nation initiative to fight against climate change | UNECE. https://unece.org/media/press/368739 22 May 2024.
3. Ministry of Ecology, Environmental Protection and Climate Change of the Republic of Uzbekistan. (2023). National state of the environment report: Uzbekistan. International Institute for Sustainable Development.
4. Ilhomjon Aslanov, Uzbekkhon Mukhtorov, Rahimjon Mahsudov, Umida Makhmudova, Saida Alimova, Lobar Vohidovna Djurayeva and Ibragimov, O. (2021). Applying remote sensing techniques to monitor green areas in Tashkent Uzbekistan. E3S Web Conf. 2021 Ural Environmental Science Forum 'Sustainable Development of Industrial Region' (UESF-2021), 258(04012), pp.04012–04012. doi: https://doi.org/10.1051/e3sconf/202125804012.
5. UNDP (2023). Uzbekistan will align its national biodiversity policy to Kunming-Montreal Global Biodiversity Framework. [online] UNDP. Available at: https://www.undp.org/uzbekistan/press-releases/uzbekistan-will-align-its-national-biodiversity-policy-kunming-montreal-global-biodiversity-framework [Accessed 20 May 2024].
6. Agostini, P., Leow, W. and Mantovanelli, M. (2023). Uzbekistan's Transition to a Green Economy: Challenges and Opportunities. [online] thediplomat.com. Available at: https://thediplomat.com/2023/02/uzbekistans-transition-to-a-green-economy-challenges-and-opportunities/.
7. International Union for Conservation of Nature and Natural Resources (2023b). The Regional Dialogue: Scaling up Nature-based Solutions for resilient landscapes in Central Asia - events | IUCN. [online] www.iucn.org. Available at: https://www.iucn.org/events/iucn-event/regional-dialogue-scaling-nature-based-solutions-resilient-landscapes-central [Accessed 19 May 2024].
8. Ahn, Y.-J. and Juraev, Z. (2023). Green spaces in Uzbekistan: Historical heritage and challenges for urban environment. Nature-Based Solutions,

[online] 4, p.100077. doi: https://doi.org/10.1016/j.nbsj.2023.100077.

9. Başsüllü, Dr.Ç., Belen, İ. and Kaptanoğlu, E. (2023). Guidelines on the Implementation of Nature-based Solutions (NbS) to Combat the Negative Impact of Climate Change on Forestry. FAO. doi: https://doi.org/10.4060/cc7450en.

10. Shahzoda Alikhanova and Bull, J.W. (2023). Review of Nature-based Solutions in Dryland Ecosystems: The Aral Sea Case Study. Environmental Management, [online] 72(3), pp.457–472. doi: https://doi.org/10.1007/s00267-023-01822-z.

11. Veckalne, R. and Tambovceva, T. (2023). Evaluating Urban Sustainability in Uzbekistan: A Novel Formula for Empirical Analysis. Sustainability, 15(9), p.7035. doi: https://doi.org/10.3390/su15097035.

12. Anvar Mukhamedjanov, Dilorom Isamukhamedova and Tang, B.-S. (2024). Green Spaces for Summer Cooling: Case Study of Tashkent, Uzbekistan. International review for spatial planning and sustainable development, 12(2), pp.163–180. doi: https://doi.org/10.14246/irspsd.12.2_163.

13. International Union for Conservation of Nature and Natural Resources (2023a). Strong commitment to foster Nature-based Solutions for resilient landscapes in Central Asia - News | IUCN. [online] www.iucn.org. Available at: https://www.iucn.org/news/202306/strong-commitment-foster-nature-based-solutions-resilient-landscapes-central-asia [Accessed 20 May 2024].

14. Erdanaev, E., Kappas, M., Pulatov, A. and Klinge, M. (2015) "Short Review of Climate and Land Use change Impact on Land Degradation in Tashkent Province"., International Journal of Geoinformatics, 11(4). Available at: https://journals.sfu.ca/ijg/index.php/journal/article/view/909

15. Global Green Growth Institute (2021). Review of Strategic Plan 2015-2020: Commitments Delivered and Key Results. [online] GGGI - Global Green Growth Institute. Available at: https://gggi.org/report/review-of-strategic-plan-2015-2020-commitments-delivered-and-key-results/

16. Asian Development Bank (2023). GREEN (Growth, Resilience, Energy Efficiency, and Nature-Based) Solutions for Livable Cities. [online] www.adb.org. Available at: https://www.adb.org/projects/57040-001/main#project-pds [Accessed 20 May 2024].

17. L., P. and T., R. (2024). Biodiversity Finance Policy and Institutional Review in Uzbekistan. [online] BIOFIN. Available at: https://www.biofin.org/knowledge-product/biodiversity-finance-policy-and-institutional-review-uzbekistan.

AS_20

Coral Guardians: A Youth-Led Revival in Kayankerni Coral Cove

Context Overview

LOCATION

NBS TYPOLOGY	ECOSYSTEM TYPE
Restoration; Protection & conservation	Reefs
IMPACT SCALE	**URBAN TYPOLOGY**
Site	Outside the city with urban impacts
SOCIAL CHALLENGES ADDRESSED	**CO-BENEFITS**
Sea level rise; Biodiversity loss	Reinforcement of socio-ecological knowledge; Stronger human-nature connections

Kayankerni Coral Cove, located on Sri Lanka's eastern coast, lies within a region rich in marine biodiversity. The area experiences a tropical climate, with average temperatures ranging from 25°C to 30°C and heavy seasonal monsoons. It hosts diverse ecosystems, including coral reefs, mangroves, and seagrass beds. However, these ecosystems face mounting environmental pressures — most notably, rising sea temperatures and land-based pollution — which have led to significant coral bleaching and ecosystem degradation.

By employing ecological methods like coral nurseries and the transplantation of resilient coral species, the initiative restores reef health while enhancing biodiversity. Coupled with strong community engagement and environmental education, the project demonstrates how nature-based solutions can offer sustainable, low-tech approaches to climate adaptation, ecosystem recovery, and local capacity-building in coastal environments.

Introduction to the case study

This case exemplifies the use of nature-based solutions to revive fragile marine ecosystems through youth-led coral restoration efforts in Kayankerni Coral Cove, Sri Lanka. Over the last few decades, the coral reefs in Kayankerni Coral Cove have encountered multiple threats, largely driven by human activities and climate change. One of the primary issues is coral bleaching, which has been exacerbated by rising sea temperatures linked to climate change, resulting in significant loss of biodiversity and diminished resilience within marine ecosystems. Additionally, the marine habitats have significantly degraded marine life due to overfishing, destructive fishing practices, and coastal development, all of which have destroyed vital habitats for numerous marine species (WUR, n.d).

Furthermore, pollution from runoff associated with agricultural activities and the surrounding urban areas has contaminated coastal waters, placing further stress on coral reef systems. Compounding these challenges is the local community's limited understanding of the dynamic nature of these threats, which often limits effective conservation efforts.

Nature-based Solutions characteristics

This project utilizes a nature-based solutions approach that focuses on the restoration and rehabilitation of coral reefs through community engagement and sustainable practices. Within the Kayankerni Marine National park, Kalkudah was selected as the first site of restoration owing to its endemicity with more than 210 fish species (Youth-led Coral Restoration Initiative, n.d). Some of the key characteristics include:

1. Coral Fragment Transplantation: Local youth have been trained in the collection and transplantation of coral fragments from healthy areas to degraded sections of the reef. This method allows for the natural recovery of coral ecosystems, using resilient species such as Acropora and Pocillopora that can thrive in varying conditions.

2. Establishment of Coral Nurseries: Coral nurseries have been established in shallow waters where young coral fragments can grow for six to 12 months before being transplanted onto damaged reefs. To date, more than 1,500 coral fragments have been successfully cultivated and reintroduced to the wild.

3. Community Awareness Campaigns: The initiative includes educational workshops and community events, such as beach clean-ups and local art competitions centered on marine themes. These activities have engaged over 1,000 community members, raising awareness about the importance of coral reefs and sustainable fishing practices.

4. School Involvement: Schools have played a critical role in the project, with students participating in hands-on activities like coral gardening and monitoring reef health. The integration of marine education into the curriculum has helped foster a generation of environmentally conscious youth who are passionate about marine conservation.

5. Monitoring and Research: A rigorous monitoring program assesses the health of the transplanted corals, measuring growth rates and biodiversity changes. This collaboration with local universities and research institutions has facilitated data collection and analysis, ensuring evidence-based management practices.

The scale of implementation encompasses both local community initiatives and broader regional projects that involve multiple stakeholders, including NGOs, government agencies, and academic institutions.

Governance

The governance structure of the project is characterized by collaboration among multiple stakeholders. Local youth groups, i.e., Earthlanka volunteers played a crucial role in implementing restoration activities, fostering a strong sense of community ownership and responsibility as they engage in all phases of the project, from planning to execution. NGOs, such as the Sri Lanka Wildlife Conservation Society, and officers from Marine Environmental Authority offered technical expertise, training, and logistical support, while also facilitating connections to international conservation funds to secure essential resources. Government agencies lent their support by enacting policies that promote marine conservation and sustainable practices, including the establishment of marine protected areas (MPAs) to protect coral reefs from destructive activities. Additionally, collaborations with Wageningen University enhanced the project's scientific foundation by providing research support and technical guidance; students frequently participated in fieldwork and data collection, enriching their educational experience (Polwatta, 2023).

Funding for the project came from a variety of sources, including international grants focused on marine conservation, contributions from local businesses, and community fundraising events. The project is ongoing, with a focus on establishing long-term sustainability and resilience for both the coral reefs and the local community.

Outcome

The youth-led rehabilitation efforts have yielded several positive outcomes. Monitoring data reveal a notable 30 percent increase in coral cover within restored areas over the first two years, alongside improved biodiversity, as more than 20 fish species have returned to the rehabilitated zones. Additionally, local fishermen have reported an increase in fish catches near the restored reefs, enhancing the economic conditions for families dependent on fishing. The project has also successfully raised awareness about marine conservation; surveys show that a greater number of participants in educational programs are committed to sustainable practices and stewardship of marine resources. Furthermore, the rehabilitation has strengthened the resilience of the coral reefs against climate stressors such as bleaching events and ocean acidification, contributing to a more robust ecosystem capable of withstanding future environmental challenges (Polwatta, 2023).

However, despite these successes, the project faces ongoing challenges, including securing long-term funding for maintenance activities, managing external pressures from tourism and fishing, and ensuring sustained community engagement beyond the initial enthusiasm surrounding the initiative.

Writing: Maitreyi Koduganti
Review: Carole Alazki
Copy Editor: Veronica Olivotto
Design: Phuong Anh Bui

References:

1. Wageningen University & Research (WUR). (n.d.). Coral reef rehabilitation project in Sri Lanka wins the Nature-Based Solutions Challenge. Wageningen University & Research. Retrieved from https://www.wur.nl/en/newsarticle/coral-reef-rehabilitation-project-in-sri-lanka-wins-the-nature-based-solutions-challenge.htm

2. Youth-led Coral Restoration Initiative. (n.d.). Youth-led Coral Restoration Initiative [Video]. YouTube. Retrieved from https://youtu.be/dZlPlWi1eiA?si=QS9kig3HDPygac6M

3. Polwatta, S. (2023). How the school met the reef. Global Landscapes Forum Retrieved from https://stewards.globallandscapesforum.org/oceans/4692/how-the-school-met-the-reef

AS_21 Climate Resilience through Mangrove Afforestation

Context Overview

LOCATION 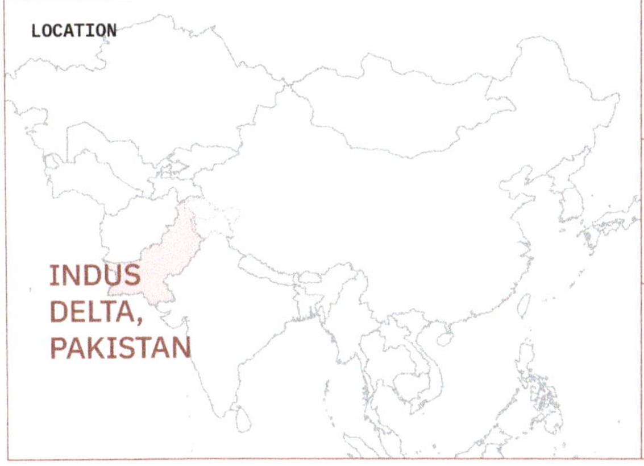	NBS TYPOLOGY Restoration	ECOSYSTEM TYPE Mangroves; Riparian areas
	IMPACT SCALE Site	URBAN TYPOLOGY Outside the city with urban impacts
	SOCIAL CHALLENGES ADDRESSED Biodiversity loss; Storms; Land degradation	CO-BENEFITS Green jobs and economic benefits; Carbon sequestration

As one of the world's largest arid mangrove ecosystems, the Indus Delta faces significant challenges, including sea intrusion, land degradation, and biodiversity loss. In response, large-scale restoration efforts — led by government agencies, NGOs, and local communities — have focused on reforestation, nursery development, and sustainable management practices. These nature-based solutions not only protect coastal habitats and reduce the risks of natural disasters such as cyclones and erosion but also contribute to carbon sequestration and livelihood generation through initiatives like the Delta Blue Carbon projects. This demonstrates how ecological restoration can effectively align with climate adaptation, biodiversity conservation, and socio-economic development goals.

Pakistan's coastline spans 1,050 kilometers, with a width ranging from 40 to 50 kilometers, covering parts of the Sindh and Balochistan provinces between 24°-25°N latitude and 61°-68°E longitude. Along this coast, mangrove forests play a vital role in climate regulation, erosion control, and coastal protection. These forests are distributed across eight distinct geographic zones, with the Indus Delta being the largest. Other significant areas include Sandspit in Sindh, and Sonmiani Khor, Kalmat Khor, Sahidi Khor, Sawar Khor, Shabi & Ankra Creeks, and Jiwani in Balochistan (see Figure 16). Globally, the Indus Delta ranks as the fifth-largest delta and the seventh-largest mangrove forest (WWF, n.d.), and it was designated a Ramsar site in 2002 (Ramsar Sites Information Service, n.d.).

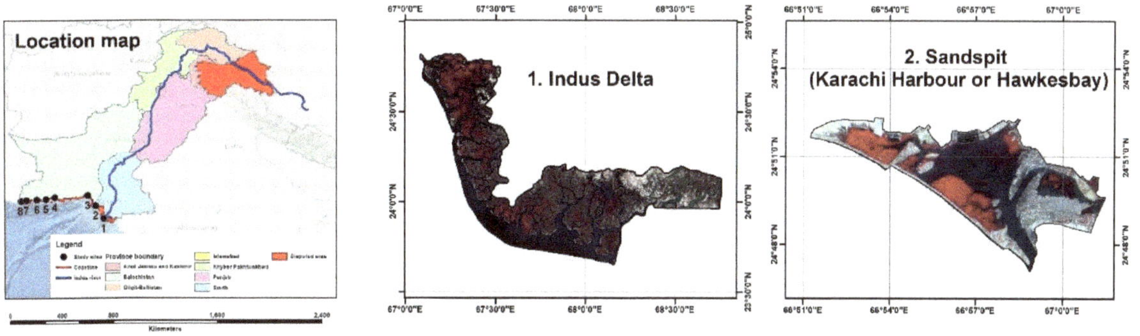

Figure 16: Eight Mangrove Sites in Pakistan (image continues on the next page).

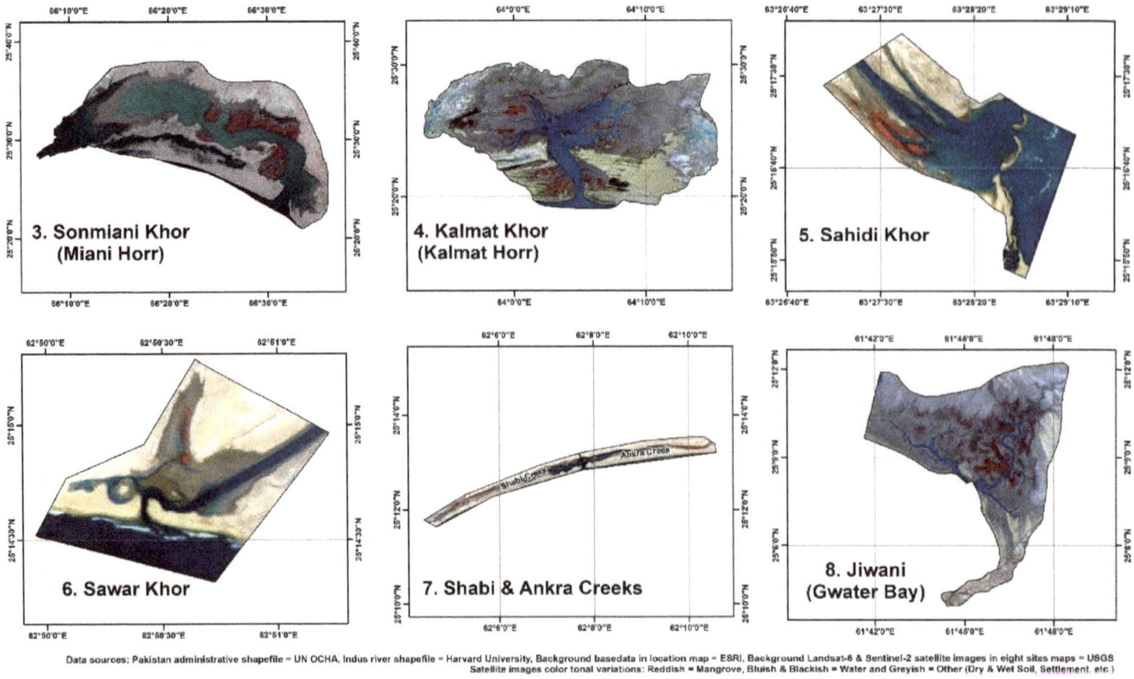

Eight Mangrove Sites in Pakistan

Introduction to the case study

This case demonstrates the application of mangrove afforestation as a nature-based solution to enhance climate resilience in the Indus Delta region of southern Pakistan. In the Indus Delta, sea intrusion (the erosion of mud flats in the creeks) is a continuous phenomenon along with soil salinity, transport, and uniform dispersal of suspended sediments. Together, these are primary causes of mangrove detrimental and degradation in the Indus delta. To reduce climate risks Pakistan national and INGOs, the Pakistan Navy, Government of Pakistan, and local Community-Based Organizations (CBOs), have been conducting mangrove plantation campaigns in coastal areas. In 2009, the Guinness Book of World Records certified Pakistan for planting 541,176 mangrove saplings in 24 hours near Keti Bandar in the Indus Delta (Dawn, 2009).

Nature-based Solutions characteristics

In 2019, as part of an ongoing campaign to improve forest cover, the Sindh Forest Department (SFD) collaborated with the World Wildlife Fund for Nature (WWF) to set up a mangrove nursery, hiring 250 women not just to plant new trees but also to guard them against threats from animals and humans. Besides employing the eco-guards, the Sindh Forest Department has given "green jobs" to over 50,000 people since 2000 — around 40 percent of them women (Kunbhar, 2021). These nature-based solutions protect the local environment and support local populations' livelihoods by enhancing biodiversity, improving coastal resilience against storms and erosion, and sequestering carbon. Pakistan's commitment to increasing mangrove cover highlights the effectiveness of integrating nature-based solutions for ecological restoration and climate change mitigation.

Governance

Back in 2016, Pakistan's first Nationally Determined Contribution (NDC) did not identify blue carbon ecosystem strategies or plans to leverage this potential (Jamshaid, 2020). In 2021, Pakistan submitted its revised NDC, including the blue carbon ecosystem as one of the new sectors. The revised NDC has an unconditional emissions reduction target of 15 percent and a conditional target of 35 percent by 2030 compared to the 2015 baseline at the national scale (WWF, n.d.b). Under the United Nations Framework Convention on Climate Change (UNFCCC), several financing mechanisms, including the Green Climate Fund (GCF), support nature-based solutions to climate change. Therefore, including these sinks in national inventories will provide Pakistan with suitable avenues for better accounting of coastal ecosystems and incentivize financial flows to coastal blue carbon interventions (Government of Pakistan, 2021).

Outcome

From 1990 to 2020, mangrove forest cover in Pakistan expanded by 300 percent (Gilani et al., 2021). Approximately 95 percent of this cover is concentrated in the Indus Delta, which spans 8,000 square kilometers and is located at 23.957°N latitude and 67.5015°E longitude (Idrees, 2022, January 30). As per official figures, Pakistan's emissions in 2018 totaled 489.87 metric tons of CO2 equivalent (MtCO2e), accounting for around 0.50 percent of global carbon dioxide emissions. The Sindh Government has been working on Delta Blue Carbon (DBC) 1 and 2 since 2015 and 2020, respectively. This suggests that the DBC-1 and DBC-2 emissions credit projects will equal approximately half of the country's emissions. Sindh is the first and only province in the country to earn 22 million USD from carbon credits. By 2075, Sindh is expected to create 140 million carbon credits in the Voluntary Carbon Market (VCM) (Korai, 2023).

Figure 17: Extensive and impressive improvement in mangrove plantations in the Indus Delta. WWF-Pakistan (n.d.)

Writing: Hammad Gilani
Review: Carole Alazki
Copy Editor: Veronica Olivotto
Design: Phuong Anh Bui

References:

1. Dawn. (2009, July 17). Guinness certifies mangrove sapling world record. https://www.dawn.com/news/478264/guinness-certifies-mangrove-sapling-world-record

2. Gilani, H., Naz, H. I., Arshad, M., Nazim, K., Akram, U., Abrar, A., & Asif, M. (2021). Evaluating mangrove conservation and sustainability through spatiotemporal (1990–2020) mangrove cover change analysis in Pakistan. Estuarine, Coastal and Shelf Science, 249, 107128. https://doi.org/10.1016/j.ecss.2020.107128

3. Government of Pakistan. (2021). Pakistan: Updated nationally determined contributions 2021. United Nations Framework Convention on Climate Change (UNFCCC). https://unfccc.int/sites/default/files/NDC/2022-06/Pakistan%20Updated%20NDC%20 2021.pdf

4. Jamshaid, S. H. (2020, November 25). Exploring the opportunity for Pakistan to include blue carbon into revised NDCs. NDC Partnership. https://ndcpartnership.org/news/exploring-opportunity-pakistan-include-blue-carbon-revised-ndcs

5. Korai, S. (2023, October 7). Sindh hopes mangrove project can balance CO_2 emissions, obtain carbon credit revenue worth billions. The Friday Times. https://thefridaytimes.com/07-Oct-2023/sindh-hopes-mangrove-project-can-balance-co2-emissions-obtain-carbon-credit-revenue-worth-billions

6. Kunbhar, Z. (2021, March 18). On Pakistan's Indus Delta, women forest rangers guard world's largest arid mangrove forest. Arab News. https://www.arabnews.com/node/1827421/pakistan

7. Idrees, A. (2022, January 30). Resilience in the Indus Delta. The Express Tribune. https://tribune.com.pk/story/2341122/resilience-in-the-indus-delta

8. Ramsar Sites Information Service. (n.d.). Annotated list of wetlands of international importance: Pakistan. Retrieved May 29, 2025, from https://rsis.ramsar.org/sites/default/files/rsiswp_search/exports/Ramsar-Sites-annotated-summary-Pakistan.pdf

9. WWF. (n.d.a). Indus Delta, Pakistan. World Wide Fund for Nature. Retrieved September 29, 2025, from https://wwf.panda.org/discover/knowledge_hub/where_we_work/indus_delta/

10. WWF-Pakistan. (n.d.b). Nationally determined contributions. Retrieved September 29, 2025, from https://www.wwfpak.org/our_work_/climate_and_energy_/nationally_determined_contributions_/

Figures by Hammad Gilani

AS_22 East Kolkata Wetlands

Context Overview

NBS TYPOLOGY	ECOSYSTEM TYPE
Restoration; Protection & conservation	Wetlands
IMPACT SCALE	**URBAN TYPOLOGY**
Greater urban region	Outside the city with urban impacts
SOCIAL CHALLENGES ADDRESSED	**CO-BENEFITS**
Flooding; Water quality; Heat; Biodiversity loss	Reduced infrastructure cost; Food security; Carbon sequestration; Air quality; Recreational benefits

Located on the peri-urban edge of Kolkata, India, the East Kolkata Wetlands (EKW) are a 12,500 hectare wetland system that naturally treats over 900 million litres of sewage daily while supporting agriculture, aquaculture, and the livelihoods of over 150,000 marginalized people. The East Kolkata Wetlands (22°25′N to 22°40′N and 88°20′E to 88°35′E), the peri-urban interface of Kolkata, is a multifunctional cornerstone ecosystem, spanning across 12,500 hectares of sewage-fed fisheries, agricultural land and solid waste farmland. It is a remnant of a series of brackish wetlands interconnected to freshwater and marine environments in an ecological continuum with the largest mangrove delta in the world – the Sundarbans. This ecosystem dates back to the period of British colonization – when through "tamed interventions" (such as excavation of canals, installation of pumps, lock-gates, and sluices), the EKW evolved as the sewage-fed wetlands, promising drainage-sewerage-utilities for the newly reclaimed colonial city of Calcutta (Mukherjee 2015, 2020).

By functioning as a sewage-fed, resource-recycling ecosystem, EKW illustrates how nature-based solutions can integrate waste management, food production, flood control, and biodiversity conservation — offering a model of circular economy and socio-ecological harmony in an urbanizing deltaic landscape.

Introduction to the case study

TheEKW serves as a living example of nature-based solutions in practice, demonstrating how ecological infrastructure can support urban resilience and sustainability. The range of challenges dotting this complex wastewater escape can be classified within broader categories of infrastructural impediments, regulatory loopholes, and inadequate awareness and mobilization initiatives. While social scientists have investigated political economic processes of wetland conversions during the contemporary times, ascribing urban sprawl as the root cause of the crisis (Dey, Samaddar, and Sen 2013; Kundu 2016), GIS investigations have validated change in land-use patterns (Momdal et al. 2022) with disruptive impacts on the coupled ecosystem services and social livelihoods.

The key issues that have been identified by the East Kolkata Wetlands Management Authority (EKWMA) and Wetlands International South Asia (WISA) from 2000–2020 are: adverse land-use change, compromised waste treatment functions, threats from invasive species, a four-fold increase of demographic density in settlement areas within wetlands, and climate risks including high floods risks, rise in temperature, and changing rainfall patterns affecting freshwater supply – the backbone of wastewater-fed fishery in EKW (EKWMA & WISA, 2021).

Nature-based Solutions characteristics

The EKW are a critical socio-ecological infrastructure, treating over 900 million litres of sewage generated by the Kolkata Municipal Corporation (KMC) every day (i.e., approximately 65 percent of the total sewage generated in the metropolitan area), saving the city nearly 4.680 million INR p.a. in sewage treatment, flood resilience services, and ensuring direct livelihood opportunities for 0.15 million marginalized people. It is often called the largest recycling ecosystem in the world, adhering to wise use of resource recovery practices, and designated as the Ramsar site since 2002 (EKWMA & WISA, 2021).

EKW is a keystone ecosystem offering an array of ecosystem services (Table 1) and benefits, and is "...a vivid illustration of more-than-human entanglements and agencies, collectively determining successful fish harvests in wastewater" (Mukherjee 2022: 42). Connected to the Bay of Bengal through the Matla-Piyali-Adiganga distributaries of the River Ganga, EKW is exposed to dynamic morphogenesis, shaped by constant interactions of fluvial, marine, and effluent discharge processes.

Provisioning Services	**Regulating**
• Fish, crop, vegetable production • Livelihoods to 0.15 million people	• Urban flood resilience efficacy - flood buffering • Carbon sequestration - climate regulation • Regulation of air pollution
Supporting Services	**Cultural**
• Biodiversity (393 plant varieties, 267 bird species, 13 mammal species, 79 fish species, 20 species of molluscs, 10 amphibians, 29 reptiles) • Groundwater recharge	• Recreational facilities • Educational components • Heritage value

Table 9: Nature-based Services offered by EKW.

Governance

A multi-tier system coordinates wastewater supply, distribution, and management in the EKW. The Kolkata Municipal Corporation pumps sewage into the wetland, while the Irrigation and Waterways Department regulates lock gates and maintains storm and sewage channels. Local water user committees allocate and monitor wastewater within individual bheris, and fishers oversee day to day pond operations in line with wise-use principles. Ownership varies across three broad categories: state owned ponds managed by the Department of Fisheries, registered or informal cooperatives run by local fishers, and a spectrum of private arrangements that range from owner managed plots to shareholding or lease based ventures. Management modes mirror these patterns, spanning state undertakings, cooperatives eligible for government inputs, joint ventures among multiple landholders, and fully commercial private companies.

The EKW Management Authority, created under the 2006 Wetlands Act, serves as the nodal agency for conservation and zoning. Between 2007 and 2013, it received 2.29 crore rupees (393,800 USD at 2013 rates) from the national Ministry of Environment, Forest and Climate Change for desilting channels, planting, and monitoring. Since 2012, the West Bengal state budget has provided 5.35 crore rupees (920,000 USD) each year for canal maintenance, removal of illegal structures, research, and outreach. A 2021 management plan under the Wetland Integrated Spatial Approach earmarks 119.8 crore rupees (20.5 million USD) for water quality improvement, habitat conservation, institutional strengthening, inventory and monitoring, capacity building, and communication. Additional support may come from the National Mission for Clean Ganga or the National Plan for Conservation of Aquatic Ecosystems, though these sources require forty percent matching funds from the state.

Writing: Jenia Mukherjee, Associate Professor, Department of Humanities and Social Sciences, Indian Institute of Technology Kharagpur, India
Review: Carole Alazki
Copy Editor: Veronica Olivotto
Design: Phuong Anh Bui

References:

1. Bunting, S. W., Kundu, N., & Mukherjee, M. (2002). Situation Analysis of Production Systems and Natural Resources in Peri-urban Kolkata. Stirling: University of Stirling.
2. Bunting, S. W., Pretty, J. N., & Edwards, P. (2010). Wastewater-fed Aquaculture in the East Kolkata Wetlands, India: Anachronism or Archetype for Resilient Ecocultures? Reviews in Aquaculture 2, 138–153.
3. Dey, I., Samaddar, R., & Sen, S. (2013). Beyond Kolkata: Rajarhat and the Dystopia of Urban Imagination. London: Routledge.
4. EKWMA and WISA (2021). East Kolkata Wetlands: Management Action Plan 2021 – 2026. East Kolkata Wetlands Management Authority and Wetlands International South Asia.
5. Kundu, R. (2016). "Their houses on our lands": Perforations and blockades in the planning of New Town Rajarhat, Kolkata. In A. Datta & A. Shaban (Eds.), Mega-urbanization in the Global South: Fast Cities and New Urban Utopias of the Postcolonial State. Oxon: Routledge.
6. Mondal, B.K., Kumari, S., Ghosh, A., Mishra, P.K. (2022). Transformation and risk assessment of the East Kolkata Wetlands (India) using fuzzy MCDM method and geospatial technology. Geography and Sustainability 3(3): 191–203.
7. Mukherjee, J. (2015). Sustainable flows between Kolkata and its peri-urban interface. In A. Allen, M. Swilling, & A. Lampis (Eds.), Untamed Urbanisms. London: Routledge.
8. Mukherjee, J. (2020). Blue Infrastructures: Natural History, Political Ecology and Urban Development in Kolkata. Singapore: Springer Nature.
9. Mukherjee, J. (2022). "Living systems infrastructure" of Kolkata: exploring co-production of urban nature using historical urban political ecology (HUPE). Environment and Urbanization 34(1), 32–51.
10. Mukherjee, J., & Ghosh, A. (2015). Water Justice City Profile: Kolkata, India. The Bartlett Development Planning Unit, University College London.

AS_23

Improving degraded riverine areas in Rohingya camps of Ukhiya, Bangladesh

Context Overview

NBS TYPOLOGY	ECOSYSTEM TYPE
Restoration; Protection & conservation; Constructed & novel ecosystems	Rivers & waterways; Riparian areas; Forests & trees; Wetlands
IMPACT SCALE	**URBAN TYPOLOGY**
Neighborhood/community	Outside the city with urban impacts
SOCIAL CHALLENGES ADDRESSED	**CO-BENEFITS**
Water security; Water quality; Social inclusion	Food security; Increased biodiversity and their functions; Flooding

Ukhiya lies in the Cox's Bazar district of southeastern Bangladesh, bordered by the Bay of Bengal to the west, and Ramu upazila, Bandarban district, and Teknaf upazila to the north, east, and south, respectively. This coastal highland, consisting of hillocks and valleys, spans 261.8 square kilometers and receives high average annual rainfall of up to 4200 millimeters (Zaman et al. 2020; Fernandes et al. 2022). The region supports tropical evergreen and semi-evergreen Dipterocarp forests and lies within the ecologically critical Teknaf peninsula — an interface between the Indo-Himalayan and Indo-Malayan eco-regions — home to diverse habitats and key species such as the Asian elephant (Hassan et al. 2018). Livelihoods in Ukhiya's 54 villages largely depend on agriculture, labor, and fishing, and the area ranks among the poorest in the district. In response to the humanitarian crisis, the Bangladesh government allocated 3,000 hectares of forestland in Ukhiya to develop settlements for Rohingya refugees fleeing persecution in neighboring Myanmar (Fernandes et al. 2022).

Although situated in a rural setting, the unprecedented scale and density of the refugee settlements — accommodating hundreds of thousands of displaced people — have imposed urban-like pressures on land, water, and ecological systems. This has made Ukhiya functionally similar to an urbanizing environment, where high population density, infrastructural demands, and ecological degradation converge. In response to rapid deforestation, habitat loss, and increased disaster risks, a series of nature-based solutions interventions were implemented, including riparian afforestation, stream restoration, and decentralized wastewater treatment. These efforts demonstrate how ecosystem-based approaches can effectively address environmental challenges typical of urbanizing contexts by restoring critical ecosystem services, improving water and food security, enhancing biodiversity, and reducing vulnerability in high-density, crisis-affected areas.

Introduction to the case study

This case presents the use of ecosystem restoration as a nature-based solution to rehabilitate degraded riverine areas surrounding the Rohingya refugee camps in Ukhiya, located in the Cox's Bazar district of southeastern Bangladesh. Owing to its location and geology, Cox's Bazar district is highly prone to natural disasters including cyclones, floods and landslides (Alam et al. 2019; Fernandes et al. 2022; Norwegian Refugee Council report, 2023). It has been identified as a moderate-risk seismic zone (Humanitarian Coordination Task Team report, 2022). The large influx of refugees has exacerbated food insecurity, unemployment and poverty in the area, while also increasing pressures on the natural resources due to intensified extraction of natural resources such as wood, fish, wild food plants, grasses and bamboo (Fernandes et al. 2022). Estimates suggest the degradation of over 5000 hectares of forests due to the construction and domestic activities in the camps (Karim and Zhang, 2021) causing a massive loss in carbon sequestration potential (Qader et al. 2020). Critical ecological processes in the landscape such as hydrology and groundwater recharge have also been impacted due to the encroachment and obstruction of freshwater streams for construction. These have increased water scarcity, pollution, stream degradation and the associated risks of landslides, erratic rainfall, flooding, high temperatures and loss of crops and livelihood assets. The camp sites often encroach or are in close proximity of wildlife movement corridors, particularly those of Asian elephants, which has increased habitat fragmentation and incidences of human-wildlife conflict (Hassan et al. 2018).

Nature-based Solutions characteristics

Ecosystem-restoration in the Kalam chhara watershed was implemented near two refugee camps in Ukhiya (camp 4 and extension, Kutupalong Mega camp), considering the natural physical landscape and available resources (Fernandes et al. 2022). The interventions included stream channel restoration, stream bank protection through riparian afforestation/reforestation, construction of water reservoirs and silt traps, wastewater treatment, conservation agriculture, raising awareness and engaging the stakeholders in stream management. The stream channels were deepened and widened to improve flow, and reservoirs and silt traps were constructed along their course to improve water quality and availability throughout the year. Native plants (e.g., vetiver, bhadails, broom grass, bamboo, Palmyra palm) were used to enhance and restore green cover and to stabilize the stream banks. The riparian zone was restored with mixed-species vegetation. Wastewater treatment was done using different approaches, including bacterial treatment, constructed wetlands and plant-based treatment using Canna sp., and the treated water was used for irrigation. Locally sourced materials were used for construction, and nurseries were established to supply the plants used in the restoration activities. Environmental education was imparted to camp inhabitants to improve stakeholder engagement (see Fernandes et al. 2022).

Governance

The Centre for Natural Resource Studies (CNRS) and the UN Refugee Centre (UNHCR) conducted a rapid assessment in the area, following which CNRS designed and implemented the nature-based solutions interventions. Experts from various disciplines, government departments and universities were consulted regularly to ensure the interventions were technically appropriate and conforming to existing policy and regulatory frameworks (e.g., Forest act 1927, Ecologically Critical Area Management Rules 2016; Bangladesh Biodiversity Act 2017 etc.). The project was carried out between 2019-2021 and incurred a cost of 441,396 USD funded by UNHCR (see Fernandes et al. 2022).

Outcome

The implemented nature-based solutions have led to improved water and food security, recolonization of wildlife in the riparian forests, restoration of key ecosystem services and reduced disaster risk. The stream channel restoration and construction of reservoirs have improved water retention, thereby reducing the risks of flooding in addition to providing habitat and resources for wildlife, domestic use, irrigation and groundwater recharge. The use of green cover for bank stabilization and construction of silt traps have reduced erosion, and in combination with the systems for wastewater treatment, these have improved the water quality and availability of usable water. Surveys revealed an increase in biodiversity of key taxa in the area, including reptiles, amphibians and mammals. Narratives from the community have revealed restoration of important ecosystem services in the area, including improved micro-climate regulation, availability and quality of food and water, and improved perception of safety from disasters. Alternative livelihoods and dietary sources such as fisheries, cultivation of vegetables and fruits are also among the key outcomes (see Fernandes et al. 2022).

Writing: Mihir R. Kulkarni (Indian Institute for Human Settlement)
Review: Carole Alazki
Copy Editor: Veronica Olivotto
Design: Phuong Anh Bui

References:

1. Fernandes R., Panwar V., Sen M. (2022). Improving degraded riverine areas in Rohingya Camps of Ukhiya, Cox's Bazar. In: Nature-based Solutions for urban climate resilience in South Asia: Cases from Bangladesh, India and Nepal. Accessed at: https://cdkn.org/sites/default/files/2022-11/NbS%20Compendium_Nov%202022_final_web.pdf on 17 Jan 2024

2. Hassan M.M., Smith A.C., Walker K., Rahman M.K., Southworth J. (2021). Rohingya Refugee Crisis and Forest Cover Change in Teknaf, Bangladesh. Remote Sensing. 10, 689. doi:10.3390/rs10050689

3. Humanitarian Coordination Task Team. (2022). Multi-hazard Response Plan for Rohingya Refugees in Cox's Bazar District 2022: Anticipatory Action and Response. Accessed at : https://reliefweb.int/report/bangladesh/multi-hazard-response-plan-rohingya-refugees-coxs-bazar-district-2022-anticipatory-action-response on 17 January 2024.

4. Karim, M.F., Zhang, X. (2021). Analysis of Vegetative Cover Vulnerability in Rohingya Refugee Camps of Bangladesh Utilizing Landsat and Per Capita Greening Area (PCGA) Datasets. Remote Sensing. 13, 4922. https://doi.org/10.3390/rs13234922

5. Norwegian Refugee Council. (2023). Strengthening Environmental Screening Capacity of Humanitarian Organizations - Environmental Screening Report on Rohingya refugee camps, Cox's Bazar, Bangladesh. Accessed at : https://eecentre.org/wp-content/uploads/2023/08/BANGLADESH-COXS-BAZAR-ENV-SCREENING-REPORT.pdf on 17 Jan 2024.

6. Quader, M., Dey, H., Malak, M. & Sajib, A. (2020). Rohingya refugee flooding and changes of the physical and social landscape in Ukhiya, Bangladesh. Environment, Development and Sustainability. https://doi.org/10.1007/s10668-020-00792-0

AS_24

Samudram Women's Federation: Women-led Turtle Conservation

Context Overview

NBS TYPOLOGY	ECOSYSTEM TYPE
Protection & conservation	Rivers & waterways
IMPACT SCALE	**URBAN TYPOLOGY**
Building	Outside the city with urban impacts
SOCIAL CHALLENGES ADDRESSED	**CO-BENEFITS**
Economic vulnerability; Biodiversity loss	Carbon sequestration; Community engagement & social cohesion; Stronger human-nature connections

The case is situated in Ganjam district, in the state of Odisha in eastern India. Odisha has a coastline stretching 482 kilometers, with several major rivers draining into the Bay of Bengal. As a result, the region supports rich estuarine and marine fisheries. In Ganjam, fishing is a key livelihood activity for rural communities, conducted at the mouth of the Rushikulya River and in the shallow coastal waters of the Bay.

While the Samudram Women's Federation operates in a rural coastal context, it addresses challenges that are inherently connected to urban systems. The livelihoods of small-scale fishers in this region are closely linked to urban markets through the supply of dried fish, pickles, and other value-added marine products. At the same time, their ecological stewardship—through activities like seaweed cultivation and olive ridley turtle conservation—contributes to the health of coastal ecosystems that offer critical services to nearby urban centers, including climate regulation, biodiversity, and food security. By building decentralized systems for income generation, ecological restoration, and equitable market access, this nature-based solutions initiative exemplifies the vital role that coastal communities play in enhancing the resilience and sustainability of urban regions that depend on functioning marine and estuarine ecosystems.

Introduction to the case study

This case illustrates how a women-led initiative in Ganjam district, Odisha, integrates nature-based solutions into coastal ecosystem management to support both conservation and sustainable livelihoods. Ganjam has mainly small-scale fishers who fish in the shallow coastal waters of the Bay of Bengal. One of the main problems they face is the reduced availability of fish: due to the rise in sea surface temperature, fish are migrating to deeper water, further out to sea and hence are out of reach of these small-scale fishers. A second challenge they face is the fishing restrictions that come into force during the breeding season of the marine olive ridley turtles. These turtles are a conserved species and hence during the peak nesting season (November to May), estuarine fishing at the Rushikulya river as well as marine fishing along the coast is restricted. Due to these factors, fishers' incomes were greatly reduced, and many became indebted to traders and moneylenders (Zachariah-Chaligne, n.d).

Nature-based Solutions characteristics

The nature-based solution in this case study involves an interlinked set of activities at the regional scale. It is embedded in the coastal/marine ecosystem. The main focus is on income diversification for fisher women. This is achieved by training them to undertake secondary processing (to make dried fish and pickles) and engage in alternative livelihoods (such as seaweed cultivation and crab fattening). Alongside, Samudram also conducts awareness programs to improve financial literacy, emphasize the importance of sustainable practices, and educate small-scale fishers on the legal frameworks that protect their rights. Further, it provides additional support such as storage facilities and helping the women negotiate for improved and equitable market access to sell their products.

Overall this nature-based solution has a sustainable development approach for the following reasons:

- it improves the incomes of marine fishers without increasing the quantum of fishing
- it creates awareness on sustainable fishing techniques and laws that support the same
- it promotes carbon sequestration (through seaweed cultivation)
- it facilitates the conservation of olive ridley turtles (by providing alternative livelihood during the fishing ban)

Governance

To address the problem of declining incomes, the Samudram Women's Federation was formed in 1993. It involved the consolidation of women's self-help groups and collective mobilization and bargaining on their behalf. Now it has expanded from Ganjam to Puri (a neighboring coastal district) and has over 150 groups with almost 6000 members. Samudram works closely with numerous development as well as conservation NGOs and fisher trade unions in these districts. E.g.: United Artists Association, Orissa Marine Resources Conservation Consortium, World Wildlife Fund-India etc. It also collaborates with government institutions such as the Odisha Forest and Fisheries Departments, Central Marine Fisheries Research Institute and Berhampur University. It has received financial support from various government schemes as well as international recognition in the form of the UNDP's Equator Prize (UNDP, 2010).

Outcome

One of the most significant gains has been the sense of empowerment felt by the women. The collective bargaining power of Samudram has enabled them to negotiate for better terms with wholesalers, exporters, traders etc. and improve their financial prospects despite existing class, caste and gender barriers. Hence this is a key co-benefit. Samudram has also encouraged its members to spend a portion of their earnings on improving access to education, healthcare and local infrastructure that will benefit the entire community. Therefore, there has been a ripple effect in the larger community as well.

Furthermore, they have also been involved in piloting other innovative programmes on the Odisha coast due to their extensive reach. For example, with the help of the Central Marine Fisheries Research Institute, the federation tested the embedding and use of artificial cement reefs – 5 villages were involved in this initiative to improve small-scale fisheries by creating additional habitat for coastal fish species. Similarly, Samudram has piloted schemes offered by the Department of Animal Husbandry, which aim to provide alternative livelihoods for fishers during the monsoon and turtle breeding seasons by providing training and support for keeping ducks, chicken, goats etc. In other words, the federation is able to funnel various forms of government support to its members.

Some of the challenges to this nature-based solution are the ability to sustain its networks given the rate of growth it is experiencing (8-10 percent per annum). It also plans to convert into a multi-purpose cooperative society to attract more financial support from banks, private sources and expand its sales to distant markets. However, this requires nurturing strong local leadership with the ability to draw up detailed financial plans and identify suitable external partners. Another point of concern is that the environmental sustainability of its initiatives need regular monitoring so that negative impacts can be addressed early on. For instance, it is not clear how seaweed cultivation (although it is important for carbon sequestration) affects the coastal ecosystem because the cultivated species are non-native ones (Zachariah-Chaligne, n.d).

Nevertheless, Samudram remains an inspiring, replicable model and it has received support from the state government to share its experiences with five other regions. It has also provided inputs to 10 other states in India via an Oxfam program on coastal ecosystems and livelihoods.

Writing: Madhuri Ramesh, School of Development, Azim Premji University, Bengaluru, India.
Review: Carole Alazki
Copy Editor: Veronica Olivotto
Design: Phuong Anh Bui

References:

1. Zachariah-Chaligne, A. (n.d). The role of Samudram Women's Federation in reducing poverty and protecting resources. 3pp. Available at: https://www.communityconservation.net/odisha-india/

2. United Nations Development Programme. 2012. Samudram Women's Federation, India. Equator Initiative Case Study Series. New York, NY.

AS_25 Oyster Reefs as Guardians of Coastal Bangladesh

Context Overview

LOCATION

NBS TYPOLOGY Restoration	ECOSYSTEM TYPE Reefs
IMPACT SCALE Site	URBAN TYPOLOGY Outside the city with urban impacts
SOCIAL CHALLENGES ADDRESSED Biodiversity loss; Sea level rise; Land degradation	CO-BENEFITS Water quality

Bangladesh is among the most climate-vulnerable countries globally, with its coastal areas increasingly exposed to sea-level rise, storm surges, cyclones, and coastal erosion. Kutubdia, surrounded by the Bay of Bengal on three sides and a river channel on the fourth, is an ecologically fragile deltaic zone facing severe shoreline loss, particularly during the monsoon season. Traditional grey infrastructure — such as earthen and concrete embankments — has proven inadequate in this dynamic, sediment-rich coastal system, often failing due to high wave energy and subsequent sediment displacement.

Introduction to the case study

This case highlights the use of oyster reef barriers as a nature-based solution to address shoreline erosion, biodiversity loss, and escalating climate risks along the eastern coast of Bangladesh, specifically in Kutubdia Upazila, an island in Cox's Bazar district. The implementation of oyster reef barriers offers a soft-engineering, ecological alternative that not only dissipates wave energy but also traps sediment and promotes the regeneration of mudflats — critical habitats supporting diverse flora and fauna. These structures provide co-benefits such as enhancing biodiversity, mitigating the impacts of overfishing and pollution, and improving coastal resilience. Declines in local aquatic, terrestrial, and avian species — exacerbated by habitat degradation and climate pressures — further highlight the need for such interventions.

Although Kutubdia itself is a rural island, the protective and ecological services provided by the reef have direct implications for nearby urban and peri-urban centers that depend on coastal stability for livelihoods, food systems, and disaster risk reduction. The initiative demonstrates how nature-based solutions can outperform conventional hard infrastructure by simultaneously addressing ecological degradation, climate vulnerability, and socio-economic insecurity. It underscores the importance of ecosystem-based approaches in building resilience across rapidly urbanizing coastal landscapes that rely on healthy marine and estuarine systems for their sustainability and development.

Nature-based Solutions characteristics

The Oyster Reef Barriers were introduced as a part of a PhD Project in Kutubdia Island in Bangladesh in 2016. Oyster reefs act as a natural break water, reducing wave energy and protecting shorelines from erosion. Globally over 85 percent of the natural oyster reefs have disappeared (Malmquist, 2011), necessitating the development of artificial reefs. In Bangladesh, artificial reefs, made up of concrete cement and shaped like a ring, were installed around Kutubdia island. Cement rings colonized with oyster larvae took around two years to grow given the environmental condition of the island (Chowdhury, 2021). The reefs altered wave patterns, reducing wave energy, and thus reducing erosion particularly during the monsoon season (Chowdhury et al., 2020). The artificial reefs also help to trap sediments and macro nutrients, building and protecting important habitats such as coastal mudflats and creating spaces for microorganisms which support a diversified small/prey fish (Chowdhury et al., 2020). This attracts larger predatory fish and other terrestrial and avian predators indicating its ability to restore important coastal ecosystems. This unique eco-engineering solution is hailed as a nature-based solution which focuses on conservation and restoration of coastal ecosystems and also helps with land development along mudflats.

Governance

The main actors involved in this project were Wageningen University of Netherlands, Royal Huskoning DHV, Netherlands and the University of Chittagong in Bangladesh. The project was led by former PhD Student and now associate professor of Institute of Marine Science in University of Chittagong, Dr MSN Chowdhury. After the successful implementation of the project, it has been accepted as a key strategy for sustainable shoreline erosion management based on eco-or bioengineering under the National Adaptation Plan of Bangladesh (2023-2050). Oyster reefs along with other shoreline eco-engineering nature-based solutions methods have also been accepted as strategy within the Bangladesh Delta Plan 2100. Within the NAP, a budget of 755 billion BDT has been proposed for upscaling eco-engineering solutions including oyster reefs. The main institution who leads this process now is Bangladesh Water Development Board, while University of Chittagong continues to provide technical support. There is however no concrete policy or framework regarding nature-based solutions approaches in Bangladesh.

Outcome

The outcomes of the coastal oyster reef sites in Bangladesh are promising and positive. An assessment of the pilot site revealed that oyster reef barriers were able to retain 54 percent of deposited sediments during the monsoon season compared to the control sites, thus protecting shorelines and helping create mudflats (Chowdhury et al., 2019). This also expanded seaward salt marshes, thus reclaiming land from the sea and indicating potential benefits to combat sea level rise. The oyster reefs had a significant positive impact on restoring mudflat communities. Compared to areas not protected by oyster reef, it showed a higher abundance of fish and macrobenthic communities. Studies also reveal a higher and distinctive diversity of faunal communities (Hynes et al., 2022). The oyster reef provides an important shelter for small and transient fish, which is highly important for local biodiversity. In the study site, juvenile fish populations increased by 114 percent while shrimp and crab populations also increased significantly (Chowdhury, 2020).

There are certain limitations to oyster reef barriers. Firstly, the reef is threatened by heavy sediment deposition, which can sink the reef very quickly. Siltation can also negatively impact colonization of oysters on the rings. Given the high rate of sedimentation across many parts of coastal Bangladesh, finding suitable sites remains a challenge. Secondly, there is still no standard process or methods for this which can be implemented at a wider scale. Finally, oyster reefs are a very new concept to local people in Bangladesh, and they are not aware about its potential benefits. Given high population density and varying feasibility of reef barriers, any such nature-based solutions must be cognizant of possible land use conflict in sites of implementation.

Writing: Jahin Sakkhar, University of East Anglia
Review: Carole Alazki
Copy Editor: Veronica Olivotto
Design: Phuong Anh Bui

References:

1. Rahman, M. C. et al. Climate change and risk scenario in Bangladesh. Asia-Pacific Journal of Regional Science 7, 381–404 (2023).
2. The Business Standard. Why artificial oyster reefs are the answer to our coastal embankments problems. (2022).
3. Chowdhury, M. S. N. et al. Ecological engineering with oysters enhances coastal resilience efforts. Ecol Eng 169, 106320 (2021).
4. Chowdhury, M. S. N., Hossain, M. S., Ysebaert, T. & Smaal, A. C. Do oyster breakwater reefs facilitate benthic and fish fauna in a dynamic subtropical environment? Ecol Eng 142, 105635 (2020).
5. Chowdhury, M. S. N. et al. Oyster breakwater reefs promote adjacent mudflat stability and salt marsh growth in a monsoon dominated subtropical coast. Sci Rep 9, 8549 (2019).
6. Hynes, S., Burger, R., Tudella, J., Norton, D. & Chen, W. Estimating the costs and benefits of protecting a coastal amenity from climate change-related hazards: Nature based solutions via oyster reef restoration versus grey infrastructure. Ecological Economics 194, 107349 (2022).
7. Malmquist,D. (2011) Loss of oyster reefs a global problem, but one with solutions. Virginia Institute of Marine Science (VIMS) https://www.vims.edu/newsandevents/topstories/archives/2011/oyster_reef_global.php

AS_26 Korea-Mongolia Greenbelt Reforestation Project

Context Overview

LOCATION

NBS TYPOLOGY	ECOSYSTEM TYPE
Restoration	Green spaces (gen.)
IMPACT SCALE	**URBAN TYPOLOGY**
Multi-national	Metropolitan
SOCIAL CHALLENGES ADDRESSED	**CO-BENEFITS**
Land degradation; Storm; Drought	Carbon sequestration; Stronger human-nature connections; Recreation-al benefits; Green jobs and economic benefits; Reduced infrastructure cost

The Korea-Mongolia Greenbelt Reforestation project spans through the middle of the golden Gobi Desert, which covers an area of approximately 500,000 square miles (1.29 million square kilometers) and is characterized by its diverse terrain, including vast stretches of sand dunes, rocky mountains, and flat plains (Ganbat, 2022). The Gobi Desert is a climate hotspot with increased aridity and desert steppe vegetation (European Environment Agency, 2023), where there is evidence that the desert's climate has changed significantly during the recent decades. Despite its harsh climate and difficult terrain, it has been providing a cultural landscape to people that influenced inner Asian identities into the modern day. Climate change and human activity have recently threatened the Gobi's natural environment, leading to several social, environmental, economic, and cultural challenges (McCarthy et al., 2022).

Introduction to the case study

The Korea-Mongolia Greenbelt Reforestation project provides insights into how nature-based solutions can be implemented transnationally, bridging urban areas (Ulaanbaatar, the capital of Mongolia) and vast ecosystems, to address key societal challenges, such as ecological degradation and sandstorms. The Mongolian Gobi Desert has been expanding rapidly due to overgrazing and climate change, triggering frequent and devastating dust storms with negative environmental, social, and economic consequences. Around 76 percent to 90 percent of Mongolia's land including Gobi Desert has been impacted by desertification and land degradation.

Desertification, caused by factors such as overgrazing of livestock and a decline in precipitation amount due to climate change, is accelerating the spread of the Gobi Desert into neighboring steppe and grassland ecosystems, making it the fastest-growing desert in the world (McCarthy et al., 2022). As a result, dust storms in Mongolia's extensive Gobi and desert-steppe zones occur frequently and cause serious disasters. In May 2008, a severe dust storm in eastern Mongolia led to the deaths of 52 people (including 14 children) and caused significant damage to buildings and infrastructure (Jugder et al., 2011). Deteriorating environmental and economic conditions have caused many people to be displaced and forced to live in urban slums to sustain themselves and their families. Traditionally, afforestation efforts in Mongolia have lacked organized governance systems. These include a fragmented land registration system and a lack of innovation and attention to forest restoration. Challenges have also been associated with the nation's economy, such as the need for aid flows and resource shortages, which collectively contribute to the disorganization and inefficiency of Mongolia's afforestation initiatives. In 2006, a collaborative three-phase reforestation initiative, Korea-Mongolia Greenbelt Reforestation Project, started to address the issues of desertification, climate change, and yellow sand dust in the Gobi Desert.

Nature-based Solutions characteristics

The Korea-Mongolia Greenbelt Reforestation Project is rooted in nature-based solutions that aim to address environmental degradation in both countries. It integrates conservation, restoration, and development approaches by employing various techniques suited to arid conditions, focusing on a unique desert ecosystem, and implementing interventions at multiple scales. The Gobi Desert is home to diverse and unique ecosystems, including vegetation, wildlife, and geological formations. By conserving and restoring the degraded ecosystems through planting over 3,000 hectares of drought-tolerant plants like dwarf elms, salt cedars, pines, etc. (Ganbat, 2022), the project intended to demonstrate that forest restoration and tree planting are possible even in adverse conditions, particularly in deserted regions. This also aligns with the concept of biomimicry i.e., learning from and mimicking natural processes to solve human-induced challenges. The project consists of multi-phased afforestation efforts, building tree nurseries, conducting capacity-building training, urban forest development, and assessing socio-economic impact (Ganbat, 2022). There was also the installation of drip irrigation systems for watering the plants which emphasized efficient water use in arid environments and building windbreaks. During this process, deliberate planning and ongoing management were also emphasized to sustain the growth of the plantations.

Governance

The Korea-Mongolia Greenbelt reforestation project was launched by the Mongolian Ministry of Environment and Tourism (MET) and the Korean Forest Service (KFS). The project has been carried out in two provinces where the sandstorm, i.e., Asian Dust, is caused. The project was carried out in three phases. The initial phase aimed to provide backing for Mongolia's "Green Belt National Program" action plan spanning from 2007 to 2016, centering on establishing effective measures for afforestation to combat desertification and land degradation. As part of the subsequent phase of the bilateral reforestation initiative that spanned from 2017 to 2021, an urban forest known as Mongolian-Korean Friendship Park was erected over a sprawling expanse of 40 hectares near Ulaanbaatar - the capital of Mongolia. With regards to future endeavors of this undertaking, plans are underway for a third phase taking place during the time frame between 2022 and 2026, with several goals including implementing a Korean forest fire prevention system, conducting research and training on agroforestry, as well as developing healing forests in Mongolia. The project was initially sponsored by KFS at USD 9.5 million with an annual management cost of 1 million USD. Mongolia took over the afforestation land after the project's first phase was completed for management (Ganbat, 2022). Still, long-term management of forest ecological systems has been a challenge as it was not considered in the initial implementation plan (Daeseob & Gyumi, 2016).

Outcome

The Korea-Mongolia Greenbelt is regarded as a successful nature-based solutions project achieving excellent results as part of Mongolia's national Greenbelt program. The project contributed to increasing forestland to 9 percent by 2030, which aligns with Mongolia's national afforestation target. In addition to afforestation, the project included slow-growing trees that store large amounts of carbon and methane, prevent erosion on steep mountainside slopes, and act as a natural barrier against desertification. Further, forest education resulted in domestic trainees, monitoring and joint research projects, employing experts in each field, and public workshops and symposiums Ganbat, 2022). In the urban forest near Ulaanbaatar, besides 90 different types of trees and other plants, there will be a learning garden, a bike path, a playground, sports fields, and a parking lot for around 100 vehicles.

According to a great deal of feedback from respondents to socio-economic surveys, there was a significant increase in income and employment in the project location. There are, however, concerns about the cost-effectiveness of planting trees in deserted grasslands and the sustainable management of afforestation land. It highlights the need for ongoing management and planning (Daeseob & Gyumi, 2016). While the project has positively influenced the target region in general, significant challenges have occurred, such as unforeseen additional expenses and unexpected changes including increased labor costs or seedling prices. Moreover, a management cost reduction plan is needed to prepare for the project transfer to the Mongolian government.

Writing: Yeowon Kim
Review: Carole Alazki
Copy Editor: Veronica Olivotto
Design: Phuong Anh Bui

References:

1. Daeseob, L., & Gyumi, A. (2016). A way forward to sustainable international forestry cooperation: A case study of the Greenbelt Plantation project in Mongolia. Journal of Rural Development, 39 (Special Issue) Available at: https://repository.krei.re.kr/bitstream/2018.oak/21742/1/JRD39-S-06.pdf (Last Accessed: February 2025)
2. European Environment Agency. (2023). Hotspot — European Environment Agency. Available at: https://www.eea.europa.eu/help/glossary/eea-glossary/hotspot (Last Accessed: February 2025)
3. Ganbat, M. (2022). Successes in forest cooperation: Greenbelt Reforestation by the Republic of Korea and Mongolia | AFoCO. Available at: https://afocosec.org/newsroom/news/forestry-news/successes-in-forest-cooperation-greenbelt-reforestation-by-the-republic-of-korea-and-mongolia/ (Last Accessed: February 2025)
4. Jugder, D., Shinoda, M., Sugimoto, N., Matsui, I., Nishikawa, M., Park, S. U., Chun, Y. S., & Park, M. S. (2011). Spatial and temporal variations of dust concentrations in the Gobi Desert of Mongolia. Global and Planetary Change, 78(1–2), 14–22. Available at: https://doi.org/10.1016/J.GLOPLACHA.2011.05.003 (Last Accessed: February 2025)
5. McCarthy, C., Sternberg, T., Hoshino, B., Banfill, J., Enkhjargal, E., Konagaya, Y., & Phillips, S. (2022). Preserving the Gobi: Identifying potential UNESCO world heritage in Mongolia's Gobi Desert. Journal of Asia-Pacific Biodiversity, 15(4), 500–517. Available at: https://doi.org/10.1016/J.JAPB.2022.08.003 (Last Accessed: February 2025)

AS_27 Stream Restoration in Cheonggyecheon Park

Context Overview

LOCATION

NBS TYPOLOGY Restoration	**ECOSYSTEM TYPE** Rivers & waterways; Stormwater infrastructure
IMPACT SCALE Neighbourhood/community	**URBAN TYPOLOGY** Urban core
SOCIAL CHALLENGES ADDRESSED Flooding; Physical and mental health; Nature connection	**CO-BENEFITS** Water quality; Air quality; Increased biodiversity and their functions; Green jobs and economic benefits

The Cheonggyecheon stream is a significant landmark in Seoul, attracting visitors and residents alike due to its scenic beauty and recreational opportunities (Cho, 2010). It was initially a naturally formed watercourse; however, the stream underwent substantial changes over time due to human interventions and development. During 1970, major streams in Seoul, including Cheonggyecheon, were paved over to make way for roads and urban development. However, after nearly two decades, the project to demolish the elevated highway and restore the stream was started in July 2003. Flowing west to east through downtown Seoul and meeting Jungnangcheon, this park connects to the Han River and discharges to the Yellow Sea. The climate of Seoul is classified as a humid continental climate with four distinct seasons with hot summers and cold winters (E. J. Kim, 2020). The Cheonggyecheon Park is a landscape urban renewal project that involved the reconstruction of the historic stream and the creation of green space.

Introduction to the case study

Cheonggyecheon is a flagship urban nature-based solution project consisting of more than 10 kilometers of restored urban stream located in Seoul, the capital and largest city in South Korea. The conversion of Cheonggyecheon into a highway led to urbanization problems, degraded ecological characteristics, and severe impacts on the diverse fauna. Road and highway overpass contradicted natural elements of the stream, reflecting the misguided decision of city officials who chose to bury the stream during the country's rapid modernization (Lee & Jung, 2016). The city highway constructed above the course of Cheonggyecheon exacerbated severe urbanization problems, such as flooding resulting from changes in natural water flow, mental and physical health problems caused by loss of natural spaces and concentration of vehicular emissions, crimes, and illegal businesses arising from changes in urban landscaping.

In addition, the pollution produced as a result of the degraded urban stream was responsible for the decreased water quality and disruption of the natural habitat and ecological characteristics of the Cheonggyecheon stream. This led to a loss of biodiversity and disruption of ecosystems that were once supported by the natural watercourse The pollution and reduction in water quality affected the aquatic life ecosystem of the stream, including benthos, fishes, amphibians, reptiles, birds, and mammals, and the overall health of the stream (Lee et al., 2020). The goal of the park reconstruction project was to reduce the atmospheric, acoustic, and water pollution and restore the degraded ancient stream by removing the highway and turning it into a public amenity for recreation and leisure (Kim, 2020).

Nature-based Solutions characteristics

The restoration of Cheonggyecheon involved a combination of conservation, restoration, and development approaches to transform the urban stream into a natural and sustainable space. The project began with demolishing an aging piece of infrastructure, a 5.8 kilometers elevated highway, and replaced the site with a green space with a reconstruction of a historic stream at its center. It provides extensive access for people to engage with the natural landscape intertwined with the urban landscape (Kim, 2020). The restoration project utilized various techniques, such as water-sensitive urban design (WSUD), low impact development (LID) and biomimicry, to improve the urban environment by restoring natural habitats, improving water quality, promoting biodiversity, and recreating a natural stream-like ecosystem As a result, the reconstruction of the stream has enhanced species diversity and decreased the percentage of exotic species compared to the unrestored urban stream (Lee et al., 2020). The restoration aimed to recreate an urban natural stream ecosystem incorporating natural and designed elements to resist floods with a 200-year frequency and simultaneously promote the image of the city as having an origin in the surrounding natural forms of mountains and streams.

Cheonggyecheon park after the restoration, Seoul (South Korea)

Governance

The Cheonggyecheon project was initiated and led by the Seoul Metropolitan Government. The restoration process included the establishment of the Cheonggyecheon Citizens Committee as an institutional channel for citizens' participation. It also involved various actors, including a coalition of non-governmental organizations (NGOs) like Citizens' Coalition for Economic, Justice, Citizens' Movement for Environmental Justice, Green Korea United, Korea Environmental Federation, Green Transportation Movement, Citizens' Network for Cultural Reform, the Cheonggyecheon Citizens Committee, and the Cheonggyecheon Restoration Center (Kim, 2020). The Cheonggyecheon restoration is known as a successful example for its multifaceted impact on the environment, urban development, and social well-being. The completed project cost 386 million USD and was completed in 27 months of demolition and construction from 2002 to 2005.

Outcome

The outcomes of the Cheonggyecheon Park project in Seoul, South Korea are a successful example of nature-based solutions because of the transformation of the stream into a civic space, addressing urban problems such as flooding, noise, water, and air pollution, increasing crime rates, and degraded stream ecosystem. The reconstructed Cheonggyecheon has led to several environmental benefits including providing flood protection for up to a 200-year flood event and can sustain a flow rate of 118 millimeters per hour (Hwang, K.Y. 2007), an overall biodiversity increase of 639 percent between 2003 and 2008, compared with the pre-restoration era, reducing the urban heat island effect with temperatures along the stream 3.3° to 5.9°C, and reduced small-particle air pollution by 35 percent from 74 to 48 micrograms per cubic meter (Revkin, 2009).

In terms of economic benefits, the restoration project resulted in increasing the price of land by 30-50 percent for properties within 50 meters of the restoration project, business growth increased in the Cheonggyecheon area by 3.5 percent, which was double the rate of business growth in downtown Seoul; and an increase in employment rate by 0.8 percent versus a 2.6 percent decrease in downtown Seoul after completion of the project. From a social benefits point of view, ridership on buses and subways in Seoul increased by 15.1 percent and 3.3 percent, respectively, between 2003 and 2008, and a daily average of 64,000 visitors were attracted to the site, of those, 1,408 were international tourists (Kim, 2020).

Despite the benefits of the restoration project to the society and economy of downtown Seoul, there were a few challenges to the development of Cheonggyecheon Park. For example, there still are differing opinions about the project's ecological and historical authenticity. The project has been criticized for neglecting the historical context of the area and developing a tourist attraction rather than restoring a natural ecosystem. However, advocates argue that the project has successfully created a recreational and ecologically beneficial public space. The controversy surrounding the authenticity of the restored Cheonggyecheon reflects ongoing debates about the ecological and historical character of the stream, indicating a need for further discussion on the sustainability aspects of restoration (Cho, 2010). Further, the project's benefits were lower than anticipated because the restored stream was not identical to the original, leading to a lower benefit-cost ratio. In addition, as reported by Lee & Jung (2016), the cost of the project was higher than expected due to the underestimation of traffic congestion caused by restoration. Changes in adjacent areas, including the construction of high-rise buildings, density issues, increased land prices, business displacement, and traffic congestion, are some of the challenges associated with Cheonggyecheon stream restoration in Seoul.

Writing: Yeowon Kim
Review: Carole Alazki
Copy Editor: Veronica Olivotto
Design: Phuong Anh Bui

References:

1. Cho, M. R. (2010). The politics of urban nature restoration: The case of Cheonggyecheon restoration in Seoul, Korea. International Development Planning Review, 32(2), 145–165. Available at: https://doi.org/10.3828/IDPR.2010.05 (Last Accessed March 2025)

2. Hwang, K.Y. (2007). Cheonggyecheon restoration and downtown revitalization. Seoul Development Institute. Seoul, South Korea. Available at: https://www.seoulsolution.kr/en/content/seoul-urban-regeneration-cheonggyecheon-restoration-and-downtown-revitalization (Last Accessed March 2025)

3. Kim, E. J. (2020). The Historical Landscape: Evoking the Past in a Landscape for the Future in the Cheonggyecheon Reconstruction in South Korea. Humanities 2020, Vol. 9, Page 113, 9(3), 113. Available at: https://doi.org/10.3390/H9030113 (Last Accessed March 2025)

4. Kim, H., & Jung, Y. (2019). Is Cheonggyecheon sustainable? A systematic literature review of a stream restoration in Seoul, South Korea. Sustainable Cities and Society, 45, 59–69. Available at: https://doi.org/10.1016/J.SCS.2018.11.018 (Last Accessed March 2025)

5. Kim, H.S., T.G. Koh, and K.W. Kwon. (2009). The Cheonggyecheon (Stream) Restoration Project - Effects of the restoration work. Cheonggyecheon Management Team, Seoul Metropolitan Facilities Management Corporation. Seoul, South Korea. Received from Kie-Wook Kwon, July 18, 2011. (Last Accessed March 2025)

6. Lee, C. S., Lee, H., Kim, A. R., Pi, J. H., Bae, Y. J., Choi, J. K., Lee, W. S., & Moon, J. S. (2020). Ecological effects of daylighting and plant reintroduction to the Cheonggye Stream in Seoul, Korea. Ecological Engineering, 152, 105879. Available at: https://doi.org/10.1016/j.ecoleng.2020.105879 (Last Accessed March 2025)

7. Lee, M., & Jung, I. (2016). Assessment of an urban stream restoration project by cost-benefit analysis: The case of Cheonggyecheon stream in Seoul, South Korea. KSCE Journal of Civil Engineering, 20(1), 152–162. Available at: https://doi.org/10.1007/S12205-015-0633-4/METRICS (Last Accessed March 2025)

8. Revkin, A. (2009). "Peeling back pavement to expose water havens." New York Times. Retrieved from Available at: http://www.nytimes.com/2009/07/17/world/asia/17daylight.html?_r=1 (Last Accessed March 2025)

Photo by Seunghee Han.

NATURA

www.ingramcontent.com/pod-product-compliance
Ingram Content Group UK Ltd.
Pitfield, Milton Keynes, MK11 3LW, UK
UKHW060642250926
475850UK00007B/56

9 781479 850402